Outdoor Life

GUNS AND SHOOTING YEARBOOK

1988

Outdoor Life Books

Published by
Outdoor Life Books, New York

Distributed to the trade by
Stackpole Books, Harrisburg, Pennsylvania

COVER: Facilities and technical expertise by National Rifle Association. **Photograph by Randy Lamson**, Director of N.R.A. Photography

At top is the Kimber Custom Classic prototype BGR in 375 H&H caliber.

The shotgun is a new Browning for 1988, the Citori Grade 6 "Lightning" 20-gauge, 26-inch barrels with Invector Chokes.

The handgun is the controversial Glock 17 pistol used by the Austrian Army and slated for use by the Norwegian Army.

Copyright © 1987 Grolier Book Clubs, Inc.

Published by
Outdoor Life Books
Grolier Book Clubs, Inc.
380 Madison Avenue
New York, NY 10017

Distributed to the trade by
Stackpole Books
Cameron and Kelker Streets
P.O. Box 1831
Harrisburg, PA 17105

Book design: Jeff Fitschen
Editorial consultant: Robert Elman

ISBN 1–55654–025–6
ISSN 0889–0978

Manufactured in the United States of America

Contents

Preface

Back when I was a slab-footed farm boy in the hills of Tennessee I had a festering passion for guns of every type. On Saturdays, when we drove to town for the weekly shopping I made a round of every store that sold guns. There were no gunshops in those days or even sporting goods stores. Guns and ammunition, fishing tackle and baseball bats were sold in hardware stores.

The guns were always situated somewhere in the rear and more or less partitioned from the rest of the store by a waist-high barricade of shotshells. The gun racks at any time might contain anywhere from a dozen to fifty firearms, both new and used, with no two even being alike. Most of these guns were as familiar to me as the warts on my fingers and even before catching my breath (I always ran from store to store) I knew which guns had been sold and which were new additions. Each of the new guns I inspected with utmost attention; feeling the actions, stroking the varnished wood and aligning the sights on imaginary deer and bears.

When I had absorbed all the excitement and information one place had to offer, I'd dash off (pausing only to glance at the ranks of sparkling bicycles) and study the guns at another hardware. Then too, there wasn't much in the way of reading about guns then. There were precious few experts in those days and the road to experience was usually an unmarked trail.

There are many more gun experts around today and the newsstands are filled with literature that touches nearly all phases of firearms and the shooting sports. Compared to the dirth of information I suffered as a lad, we now seem to be awash in a riptide of shooting data.

That's what this yearbook is all about, the distillation of great piles of shooting lore into useful information. The process is a long one but quite simple: Throughout the year I sift through just about every shooting journal available, clipping the articles that have special merit. The collection of articles, which usually numbers well over a hundred, is then reread with each piece being accepted or rejected on the basis of several vital criteria: Most important of these is factual content, but there is still room for well-stated opinion. For example, in "Is There an Ideal Handgun?" Bob Milek and Ross Seyfried offer differing viewpoints but each draws on large reserves of personal experience.

The articles must also be well written. Fancy writing is not good enough. I look for writing that flows like cold spring water from a tin dipper, the kind of words you don't have to think about but leave you refreshed and nourished. The articles we select also have to be well illustrated. We've never had better pictures than this year or more of 'em. And speaking of art, "The Art of American Arms" by R. L. Wilson and Richard Alan Dow is an achievement in firearms photography that I've never seen surpassed. It's a wonderful bonus this year.

Also, in order to make it into this annual, an article must tell you something that's worth knowing or tell you how to do something that's worth doing. "The Glock 17 Pistol" by Pete Dickey, for instance, describes the revolutionary and controversial "plastic" handgun that has been so much in the news of late. Can it really pass through airport x-ray machines without being detected? Pete answers this question and a lot more about the Glock.

Then we leap toward tomorrow with Howard French's "Computer Programs for Shooters, Hunters and Reloaders," and my own computer-generated tables, "The Wayward Wind," that tells you how much the wind blows your bullets off course.

Another vital consideration when selecting articles for this annual is that there be something for everyone. If you are a do-it-yourself gunsmith or just enjoy making a good thing better you'll profit from "Remedy for Ailing Accuracy" by Bob Milek or Chic Blood's "Accurizing Springfield Armory's .45 Auto." Or if you just want to know what is new in the shooting world there's a roundup of the latest in firearms and shooting optics.

The list goes on: Shotguns, Handguns, Black Powder, Antiques, Rifles, Ballistics and Ammunition. Altogether there are over thirty chapters. These are the best reading the year has to offer, in the *1988 Outdoor Life Guns and Shooting Yearbook*. It's a lot better than standing around the hardware store.

Jim Carmichel

PART ONE

RIFLES

Pick a Gun With Punch

Jim Carmichel

Do you ever play the cartridge game? It's a sort of big-game trivia quiz where the players try to name all of the little calibers that they think will work for the biggest of big game. It's a great clubhouse sport because no one is called on to prove anything or even offer any firsthand experience. Of course, if a player can cite the time some famous writer/hunter downed a moose with a .22 Savage Hi-Power, it may add points to his score.

The dialogue goes something like this:

First Player: "The .300 Winchester Magnum would be great for grizzly bears."

Second Player: "Yeah, but it kicks too hard! A .270 or .30/06 has less recoil, and the ammunition is cheaper."

First Player: "But will a .270 stop a grizzly?"

Second Player: "Sure it will, I know I read it somewhere."

Third Player: "I shot a .270 once, and it kicked pretty hard. My .257 Roberts kicks less and it's great for woodchucks and deer. I even got an antelope with it year before last. Old what's-his-name, the writer, said he killed lots of moose

with a .257. A moose is bigger than a bear, ain't it? So why wouldn't a .257 be good bear medicine?"

First Player: "I dunno, never shot a .257 myself, but I hear they're okay."

Second Player: "The .257 is awfully small. I'm going to stick with my .270 for bears. The extra power may come in handy."

Third Player: "Yeah, me, too."

The problem with the cartridge trivia game is that it tends to blur the line between fact and fancy. It's one thing to theorize about the effectiveness of a particular rifle/cartridge combination but another thing entirely to be within spitting distance of an Alaskan brown bear with only a rifle and a bullet between you and kingdom come. This is when theory goes all to hell. This is when big bullets are never big enough. This is when fear of recoil becomes meaningless as you become intimate with a whole new brand of fear.

If you are fascinated by *theoretical* killing potential, you can rest assured that very large animals

This article first appeared in Hunting Guns

Which caliber you choose is a proportional proposition; the bigger the animal, the more cartridge power you need.

can be killed in their tracks with very small calibers. Vern O'Brien, for example, used to publicize his .17-caliber rifles by using them to bag big Alaskan brown bears. And in my latest book, *The Book of the Rifle*, I casually state that I'm willing to make a bet of considerable size that I can make a one-shot kill on any North American game with the lowly .22 Hornet. *But!*—there are some strings attached. O'Brien, I'm sure, picked the time and place, then placed his shots with deft precision. I would do the same with my .22 Hornet, stalking at easy range and taking dead certain aim at specific vital spots.

Though such tricks may demonstrate that big game *can* be bagged with small calibers, they do not prove a thing about what goes on in the real world of hunting. In the world of hunting, as it actually exists, there are very few opportunities to pick the time and place for a shot. You fire at game when the opportunity presents itself, and very seldom do these opportunities present themselves when circumstances are as you would like them to be. Almost never will you get the ideal shot that will allow you to prove your pet theory about your pet caliber or load—especially if your theory is that ''real'' men shoot little bullets at big game.

When you assume the role of hunter, you also assume important obligations. You are obligated, of course, to yourself; you have an obligation to the safety of your guide and hunting companions; and, most of all, you have an obligation to your quarry, to take it as quickly and cleanly as possible. If you to hedge on your equipment, you are also hedging on your obligations—which may be inexcusable, if not deadly.

There is no mystery in choosing a suitable rifle and cartridge for any big game. For the most part, it is a purely proportional proposition. The bigger the animal, the more cartridge power you need. In North America, there are relatively few species of big or dangerous animals. This situation is reflected in lists of American hunting calibers, in which there is a very small selection of suitable calibers for our biggest and meanest game.

The North American game species that I consider big and/or dangerous include the bison, moose, grizzly, polar bear, brown bear (including the Kodiak variety), musk ox, and walrus. It can be argued that elk and big caribou are in this category as well. Caribou, in my experience, do not require a lot of killing, but the longer I hunt elk, the more inclined I become to favor them with heavier calibers. This is probably because a good bull elk is getting a lot tougher to come by, and when I get a shot, I want some results. It might also be pointed out that a big black bear is as heavy as a medium-size grizzly. After visiting with these bears for a short while, however, you get the idea that the main difference between the two is more a matter of disposition than size.

Bears are tough critters. I've never seen a grizzly, and have seen only one Alaskan brown, go down in its tracks, even after being well hit with impressive artillery. Their nervous systems seem to be pretty much immune to the shock effect that signals instant lights-out for many species. I suspect that it would take a cartridge of the size and velocity of a .378 Weatherby (270-grain bullet at 3,180 feet per second) to generate enough shock effect to jerk the world out from under a big bear. And I'm not so sure that even that would work.

When bears are hit, they tend to thrash around quite a bit, eventually dropping from loss of blood. That's why I want a bullet that goes deep and stays together. A shoulder shot that breaks lots of bone before penetrating the lungs is my favorite because it keeps the bear pretty much immobile. The one brown bear I saw knocked flat had already taken a couple of close-up hits and was headed for deep cover when it caught a .338 slug at the base of the skull. Ordinarily, though, you don't shoot bears in the head, because the skull is a prized trophy.

Those impossible yarns you've heard about how a moose can soak up a boatload of bullet energy and go on munching the willows aren't at all impossible. Not long ago, my hunting pal pumped three slugs from his 7mm Remington Magnum into a moose's vitals before he got its attention. Several times, I've hit a moose hard from up close and then stood around and waited while it considered what was happening. Follow-up shots don't seem to help at all if the first bullet is in the right place. I've never had a moose get away after being hit, but it has often occurred to me that if one decided to run, it could go a long way before falling.

Because moose aren't much impressed by high-velocity bullet shock, I like to hit them with heavy, tough-jacketed bullets that will penetrate deeply. This is your best way to ensure that they won't go far.

Bison, despite their size, don't seem to be all that tough to kill. I've watched a couple of buffalo-control shoots and was impressed by how quickly they went down when shot by relatively light (.300 Magnum) rifles. But then, part of the reason they went down so quickly was because they were being culled by expert marksmen who knew exactly where to place their bullets.

I've never hunted musk oxen or walruses, so I can't offer any firsthand knowledge, but I know a few guys who have and they say that the musk ox isn't all that tough to kill. Everyone I've talked to, though, used a .300 Magnum or larger to hunt these peculiar creatures of the Far North.

A big bull walrus can be more than a dozen feet long and weigh 1½ tons, making it as bulky as a bison. Professional guides who hunt them tell me that the trick to successful hunting is getting an instant kill. Otherwise, they dive into the water and are lost. Some guides recommend the .375 Holland & Holland because of its stopping power. Others feel that a more important consideration is good accuracy and a flat trajectory because of the need for pinpoint bullet placement—provided, of course, that you can figure out *where* to place your bullet.

My threshold cartridge for any of the above species is the .30/06 with a *good* 180- or 200-grain bullet (*not* the old roundnose 220-grain slug, which can only get you into trouble). Right now, I can hear fans of the .270 and .280 scream that their favorites develop as much or more *energy* than the .30/06 and are therefore just as good. In some circumstances this might be true, especially when those cartridges' extra velocity and flatness of trajectory yield superior range—but not when it comes to boring deep into the vitals of big animals. High energy can be generated simply by boosting velocity, but when it comes to big game I want more than mathematical/theoretical energy. What's needed is the brand of knockdown power that I call "punch," that is, the power to push a bullet through bone and gristle and keep it going. Punch, I've observed, tends to be most reliable when bullets leave the muzzle at somewhere between 2,600 and 3,000 fps. If the bullet is going much faster, it may do the wrong thing on impact, such as come apart.

I am not inclined to rate the 7mm Remington Magnum as significantly better than the .30/06 for really big game. This is because of the relatively lightweight 7mm bullets currently loaded. A 200-grain slug in the 7mm Magnum might be a very interesting proposition, however. The really effective cartridges for large and/or dangerous North American game are what I call the "middle magnums." These are belted magnums of .30 caliber and larger. On the lower end of the middle magnums is the .300 Winchester, with the upper limit being the .375 H&H. In between are such solid performers as the .300 H&H, .308 Norma, .300 Weatherby, 8mm Remington, .338 Winchester, .340 Weatherby, .350 Remington, and .358 Norma. The .378 Weatherby could be included in this group, but to my notion it exceeds the need by a considerable margin and kicks like thunder.

The cartridges I've named are essentially what

Author's favorite rifles for heaviest North American big game are, from left: .338 Winchester Magnum on Mauser action; .375 Holland & Holland on Winchester Model 70 action; .358 Norma Magnum on Mauser action; and Ruger Model 77 in .350 Remington Magnum.

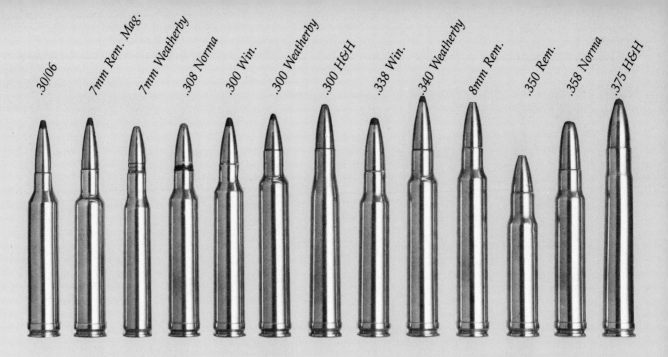

.30/06 7mm Rem. Mag. 7mm Weatherby .308 Norma .300 Win. .300 Weatherby .300 H&H .338 Win. .340 Weatherby 8mm Rem. .350 Rem. .358 Norma .375 H&H

Stanley W. Trzoniec photo

the English refer to as the "mediumbores." These were designed primarily for larger Asian and African game, including such thin-skinned dangerous game as the lion, leopard, and tiger. But whereas American cartridge makers offer only a few such calibers, British and Continental arms makers have designed them by the dozen. (The .308 and .358 Norma Magnums, though made in Sweden, were developed with North American hunters in mind.)

My pick of the middle magnums, indeed my favorite round for any of North America's big or dangerous game from elk on up, is the .338 Winchester Magnum. This is the round that offers about everything a hunter could want. These features include very nearly the punch of the .375 H&H plus a shorter case length that permits use with standard (.30/06-length) rifle actions. I always use 250-grain bullets in the .338, loaded to a muzzle velocity of slightly less than 2,700 fps. My standard load for the .338 is 68 grains of IMR 4,350 propellant behind a 250-grain Nosler Partition bullet with WW cases and CCI Mag primers.

Federal offers a superb loading with a 210-grain Nosler bullet, and Winchester offers a choice of 200- and 225-grain bullet weights. Why the Winchester people don't offer a 250-grain loading is a mystery. Perhaps they haven't looked at enough bears.

When hunting our biggest game with hypervelocity rounds such as the 7mm and .300 Weatherby Magnums, I would be very much inclined to swap velocity for bullet weight. A good 200-grain bullet, such as the Nosler Partition, in any of the .30-caliber magnums would be wonderfully effective.

RATING CALIBERS

Here is Jim Carmichael's ranking of the calibers commonly available for the bigger North American game species. His ranking is not based purely on cartridge energy but on other factors as well, including bullet and rifle availability, action length, and overall shootability.

Caliber	Score
.338 Winchester Mag.	10
.375 H&H Mag.	10
.340 Weatherby Mag.	9
.358 Norma Mag.	8
.300 Weatherby Mag.	7
.300 Winchester Mag.	6
8mm Remington Mag.	6
.350 Remington Mag.	5
.300 H&H Mag.	5
.308 Norma Mag.	5
7mm Weatherby Mag.	5
7mm Remington Mag.	4
.30/06	4

Quite a few of the Alaskan guides with whom I've hunted carry .375 H&H rifles when after big bears. They figure that their shooting will be done in a do-or-die situation and that the big bullet will pull their fat out of the fire. I agree with their rationale.

All of this talk about magnums may be frightening to the recoil-conscious, but there really isn't any way around it. If you intend to hunt bigger game, you have to use a bigger cartridge. Added recoil is one of the prices you have to pay. If you are unwilling to live with some recoil, it is best to reconsider your priorities and hunt something else.

Ultra Light Arms

Pete Dickey

THE term "custom rifle maker" has many interpretations. Used loosely, it is applied even to one who cuts down the military stock of a Mauser or Springfield rifle and produces a "sporter." At the other extreme, it could apply to one who made and assembled "lock, stock, and barrel" from scratch. If such a man ever existed, he would have had to cut his own tree and mine his own ore just to get started.

Melvin Forbes of Ultra Light Arms, Inc., comes far closer to the latter definition than most, since he makes his own locks (or actions if the first cliché annoyed you) and stocks. Yet he thinks of himself as more of a gunsmith or small "totally U.S." manufacturer than a custom gunmaker, for Douglas makes his barrels, Timney his triggers, and Pachmayr his recoil pads.

In his West Virginia plant, just across the Monongahela River from Morgantown and hard by the Pennsylvania border, Forbes now has a production capacity of five rifles a week and a four-week delivery delay. He employs three work-

This article first appeared in American Rifleman

COMPARATIVE MEASUREMENTS AND WEIGHTS

	ULA	*Rem. ADL*
Receiver length	7.50"	7.85"
Receiver diameter	1.22"	1.36"
Bolt body diameter	.594"	.696"
Bolt lug diameter	.856"	.986"
Action weight	20 ozs.	30 ozs.
Barrelled action weight	60 ozs.	76 ozs.

Melvin Forbes founded Ultra Light Arms in 1980 and then used reworked actions of other makers as the heart of his production. Now he combines his own actions and stocks with other U.S.-made components.

ers—three more than he had in 1980 when the operation began.

The basic product is called the Ultimate Mountain Rifle and is produced in four action lengths, each in right- and left-hand versions. (For prices and other details, the firm may be addressed at Box 1270, Granville, WV 26534.)

Some famous bolt-action makers have contented themselves with a single right-hand length. Others offer long and short versions. A few have gone so far as to make three lengths, but four sizes and their left- or right-hand options seem to be necessary to Forbes.

His model numbers—20S, 20, 24, and 28—indicate the weight in ounces of his actions (less trigger guards but including triggers). The Model 20S—for the .17 and .222–.223 cartridge family—weighs the same as, but is 5/8 inch shorter than, the Model 20 for such medium-length cartridges as the .22/250 Remington, .243, .308, and .358 Winchester. The ''standard-length'' Model 24 is for the .25/06 and .280 Remington, the .270 Winchester, and, of course, the .30/06. The relatively ''heavy'' Model 28 handles the 7mm Remington Magnum or Winchester's .264, .300, and .338 Mags.

Rack of ''sold'' rifles, each different and each awaiting completion, proofing and accuracy testing, represents about three weeks of production for Forbes' factory. Four-week delivery time is to be expected.

High volume of production is not immediate goal of compact Ultra Light factory located in Granville, W. Va. Firm's rugged but unsophisticated machinery and tooling accomplish the many operations necessary, such as receiver threading (left) and recoil-plate grinding.

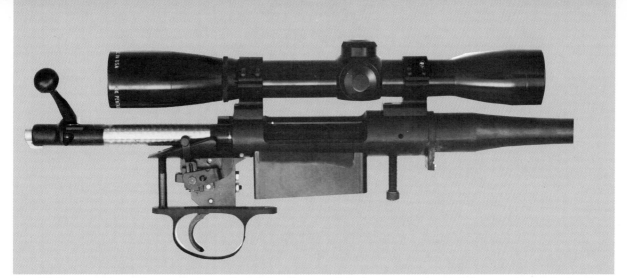

In addition to lightweight synthetic stock, ULA reduces weight by means of scaled-down action and aluminum scope mounts. Action is reminiscent of blind-magazined Remington 700 ADL, but differs in several details. Trigger is modified, fully adjustable Timney.

If enough recoil-ignorers should demand it in the future, a Model 32 *could* appear for the H&H Magnum cartridges, but, understandably, Forbes has no plans for a fifth size at this particular time.

The actions are very reminiscent of the blind-magazined Remington Model 700 ADLs, as they are based on tubular receivers (of heat-treated, stress-relieved 4140 steel) with recoil lug plates trapped between the barrel shoulder and the receiver ring. Instead of employing a bottom-activated bolt stop as in the Remington, the ULA uses a simple tab that is integral with the bolt stop itself and emerges just behind the left side of the receiver bridge.

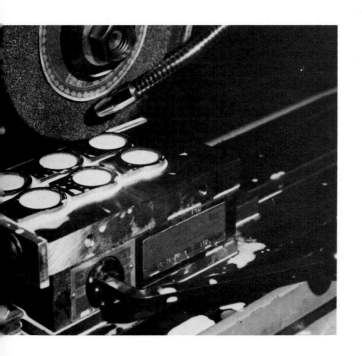

The fully adjustable Timney 700 triggers, which normally are applied as after-market options to Remington Model 700 rifles using the Remington safety parts, are modified by ULA using its own safety parts. In the rearward position, the bolt is locked, and the safety is engaged; when moved straight forward, the safety and bolt lock are disengaged. The ULA modification offers a third mode that permits bolt operation and unloading with the safety engaged. This mode is reached by simply pushing firmly down on the safety arm or button when it is moved to the safe position. The system works well, and conversion units for existing Remington rifles with factory triggers are also sold by ULA.

The 4340 steel one-piece bolt body, with dual front lugs, plunger ejector, and brazed bolt handle, also resembles the Remington, but has a small hook extractor with plunger and coil spring like that used by Sako.

A .243 Model 20 ULA rifle was examined side by side with a Remington 700 ADL in 6mm Remington caliber, in what seemed a good comparison as both used similar cartridges and had four-shot blind magazines. The 20-inch barrel of the Remington was thicker but 2 inches shorter than that of the ULA. The approximate action weights given below exclude the similar aluminum trigger guards, steel magazine parts, and recoil lugs. The barrelled actions' weights include them.

The barrelled action obviously does its share in ULA's goal of the lightweight "ultimate," but there are two other factors that help. The minor, but interesting, factor is the ULA-made aluminum scope mounts. The split rings are conventionally held with two pairs of opposing hex screws but are fixed to their solid bases which, with two screws each, are set directly on the

ACCURACY RESULTS
Five Consecutive 5-Shot Groups At 100 Yards

.243 Win. Cartridge	Vel.@15 ft. (fps)	Smallest (ins.)	Largest (ins.)	Average (ins.)
Federal Premium No. P243C 100-gr. BTSP	2,853 Avg. 24 Sd	1.03	2.23	1.67
Hornady #8040 75-gr. HP	3,212 Avg. 19 Sd	1.62	2.86	2.07
Norma No. 16003 100-gr. SP	3,042 Avg. 17 Sd	1.17	2.18	1.74
PMC 243A 80-gr. SP	3,105 Avg. 39 Sd	.96	2.30	1.64
Sd (standard deviation)		Average Extreme Spread		1.78

Exactly one-third of Ultra Light's labor force drills ejector hole in unfinished bolt of ULA Mountain Rifle.

NO GLASSES!   RKG

receiver bridge and ring. There are no intermediate seats or clamps to complicate matters or increase scope height.

The major weight reducer, after the action, is the stock. It is full-length glass-bedded from the receiver tang to the front of the barrel channel. After a great deal of experimentation with various synthetics, Forbes settled on Kevlar as his stock's primary material because of the "light weight, good tensile and excellent shear strength." For added tensile strength, graphite is used as a reinforcement, but there is no foam filler, as Forbes believes foam adds nothing but weight, slight though that might be.

The lines of the stock are "classic," with a 13¾-inch pull, and the cheekpieced comb slopes down in front to reduce felt recoil to the cheek. Molded into the stock are the sling swivel studs, a threaded insert for the short front trigger-guard screw, and a seat for the hex screw that threads into the receiver bridge. The unventilated rubber recoil pad is permanently affixed. Many colors of Du Pont Imron paint, including camouflage mixtures, are available as final finishes.

The Model 20 stock that we weighed came to 22 ounces with recoil pad and inserts installed. A similarly equipped stock of dense walnut might easily weigh twice as much.

Forbes says he could have taken the lightweight theme further by such tricks as a perforated magazine box or aluminum instead of stainless-steel follower but, he feels, these are "Mickey Mouse" approaches and could lead to trouble. Shorter barrels are possible, too. But Forbes generally sticks with his "standards" (22 inches for all save the 24-inch Model 28) for better velocity and balance.

The total weight of our .243 Model 20 sample with scope mounts came to a hair over 5 pounds.

With this to start with, plus a made-in-U.S.A. Pentax 4X scope and a Mongoose 1-ounce sling from High North Products. Antigo, Wisconsin, our rifle was rather effortlessly trundled off to the range.

The two experienced riflemen who shot the rifle agreed that it handled well and functioned perfectly. The table herewith shows that it shot well, remarkably well when one considers that with scope, rings, sling, and Michaels swivels, the entire outfit weighed an even 6 pounds.

Apologies are due to those light-rifle fanatics who have had to struggle through all this for that figure, the only point of interest to them. But, as one light-rifle fan (not fanatic) put it, "If light weight counts for everything and accuracy for nothing, the solution is no rifle."

Long-Range Varmint Rifles

Edward A. Matunas

Long-range varminting demands equipment that is up to the task. The varmint hunter interested in long-range shooting should carefully choose his rifle, cartridge, and sights.

Cartridge selection is the first important aspect of choosing a long-range varmint rifle. Cartridges such as the .22 Hornet and the .222 Remington, while extremely accurate, lack the flatness of trajectory and the long-range bullet performance necessary if shots from 250 to perhaps 400 yards are to become a reality.

The ubiquitous .223 Remington cartridge is handicapped at ranges that exceed 300 yards, as are such old favorites as the .250/3000 and .257 Roberts. And once-promising cartridges such as the .225 Winchester and .264 Winchester Magnum have not withstood the test of time.

Nonetheless, there are no fewer than nine popular cartridges that are particularly well-suited to the task of scoring clean kills on remote and tiny targets. But even among these well-qualified cartridges, there are specific advantages or disadvantages that should be considered.

The .224 Weatherby Magnum is accurate. In my Weatherby Varmintmaster, factory ammo supplies average accuracy of 1¼ inches (based on five 5-shot groups fired at 100 yards.) The best reload supplies average accuracy better than 3/4 inch. The only factory-loaded round that supplied flatter trajectory than the .224 Weatherby Magnum was the .257 Weatherby Magnum (based on velocities actually obtained with test ammo). The accuracy of the .224 with reloads ties it for first place with the .22/250 Remington among my candidate rifles.

The .224 Weatherby is a bit finicky to reload. However, Hodgdon's H335 powder seems to supply the greatest possible accuracy. Best accuracy is often obtained with 50-grain bullets. Those who prefer 55-grain bullets might better select our next cartridge listing. For specific details of the best load in each of the tested cartridges, see the accompanying table.

The .22/250 Remington is an accurate cartridge and less than m.o.a. (minute-of-angle) groups seem to be the norm rather than the exception. Indeed, it is the only caliber that, when used with factory ammunition, supplied less than 1-inch average accuracy in our tests. Therefore, for the non-reloader it represents the best possible selection for long-range varminting.

And the .22/250 is easy to load for, providing superb accuracy with a very wide range of components. The very best accuracy, however, most often comes with 55-grain bullets and Hodgdon H380 powder.

That the .220 Swift's popularity is indeed on the wane is due to several factors. It is somewhat finicky when it comes to reloads; factory ammo is not always generally available (it is made by both Hornady and Norma); the .22/250 always seems to outperform it. Yet, when matched to an accurate rifle, the Swift still gets the job done. However, after more than 30 years of varminting with and without the Swift, I feel that the .224 Weatherby Magnum and .22/250 Remington are more satisfactory .22 caliber rounds.

The .243 Winchester is the second most popular long-range cartridge today, and its popularity is well-deserved. Rifle after rifle, the .243 Winchester consistently produces m.o.a. accuracy.

In actual use, however, it is not as flat-shooting as some have dreamed it to be. Looking at the accompanying table will reveal that with factory ammo, and reloads as well, the .243's trajectory proved less flat than any of the three listed .22 caliber cartridges. The difference, however, is so small as to become nearly meaningless.

Universally, the best propellant for reloading the .243 has been IMR 4350, and our recent testing proved it.

When the varminter would like to be able to

This article first appeared in American Rifleman

SOME POPULARLY PRICED LONG-RANGE VARMINT RIFLES

Brand	Model	Available Calibers	Barrel (ins.)	Wt. (lbs.)	Comments
Remington	700 Varmint Spl.	.22/250 Rem., 6mm Rem., .243 Win.	24	9	BDL grade with heavy barrel and no sights
Remington	700 ADL	.22/250 Rem., .243 Win., .25/06 Rem., .270 Win.	22 or 24	7¼	24" bbl only for .22/250 and .25/06 Rem. cals.
Remington	700 BDL	.22/250 Rem., 6mm Rem., .243 Win., .25/06 Rem., .270 Win.	22 or 24	7¼	Available in left-hand version in .270 Win.
Ruger	77R	.22/250 Rem., .220 Swift, 6mm Rem., .243 Win., .25/06 Rem., .270 Win.	22 or 24	7	24" bbl only in .220 Swift and .25/06 cals.
Ruger	77-V Varmint	.22/250 Rem., .220 Swift, 6mm Rem., .243 Win., .25/06 Rem.	24 or 26	9	26" bbl only in .220 Swift heavy barrel model
Ruger	No. 1 Spl. Varmint	.22/250 Rem., .220 Swift, 6mm Rem., .25/06 Rem.	24	9	Heavy-barrelled single-shot model
Savage	110-V Varmint	.22/250 Rem.	26	9	Heavy barrel and special varmint stock
Weatherby	Varmintmaster	.224 Wby. Mag., .22/250 Rem.	24 or 26	7	24" bbl is med. wt. 26" bbl is heavy wt.
Weatherby	Mark V	.240 Wby. Mag., .257 Wby. Mag.	24 or 26	8	24" bbl is med. wt.; 26" bbl is heavy wt.
Browning	A-Bolt	.22/250 Rem., .243 Win., .25/06 Rem., .270 Win.	22	6½-7½	.25/06 Rem. and .270 Win., go 7½ lbs.
Mossberg	1500 Varmint	.22/250 Rem.	22	9¼	This is the old S&W varmint rifle

use his rifle for occasional antelope or deer hunting, the .243 does have this dual application ability, something no .22 caliber rifle can do. Indeed, at least one very experienced and well-known hunter and writer, Les Bowman, considers it the best deer cartridge for most shooters.

The 6mm Remington is popularly referred to as a twin of the .243 Winchester. Yet its popularity is but a small fraction of the Winchester round. Based on my experience with the 6mm, I am not surprised, as it is a bit harder to make the 6mm shoot as well as the .243. And it almost always requires more propellant to obtain .243-level ballistics.

Because rifles offered in the 6mm chambering are almost always offered in .243 Winchester, it would appear to make good sense to select the easier-to-load .243. But the reloader who is willing to put a little extra effort into load development will find the 6mm up to expectations.

Like its smaller relative, the .240 Weatherby Magnum's ammo, brass, and even rifles are, at best, less than commonplace.

With approximately 10 foot-pounds of free recoil, this is also the first of the varmint cartridges that has recoil level high enough to be considered objectionable by at least some varminters.

The tested factory loads, were, in my opinion, not accurate enough for truly long-range work. However, accuracy with reloads was satisfactory, supplying approximately 1-m.o.a. results.

The .25/06 Remington with reloads is often capable of 1-inch or smaller groups. Seldom, in my experience, do such small groups occur with factory ammo. Indeed, the largest test group average results occurred with .25/06 factory ammo.

Nonetheless, a reloader who doesn't mind some recoil will have no difficulty in obtaining satisfactory results. However, based on accuracy, the .243 Winchester cartridge would be preferred by most, albeit the .25/06 will get to the distant target with a little flatter trajectory and it will prove a better deer and antelope cartridge.

The .257 Weatherby Magnum is another cartridge that requires handloads in order to obtain sufficient accuracy for varminting. Its recoil, at

14 foot-pounds (based on a hypothetical firearm weight of 9 pounds), is the heaviest of all the varmint loads listed in the accompanying table. Accuracy is almost identical to the .240 Weatherby cartridge.

With its comparatively heavy powder charge, this cartridge would provide the shortest barrel life of the calibers for long-range varminting.

The .270 Winchester, despite its large bore diameter, is suitable as a primary selection for varmints. While its universal application has been for big game, it supplies quite a flat trajectory and plenty of long-range bullet performance.

Less-than-m.o.a. groups are not hard to come by when good handloads are used. However, based on accuracy, factory-load varmint-weight ammunition is not suitable for more long range.

Our test .270 supplied the third best handload accuracy, being surpassed only by reloads in the .224 Weatherby Magnum and the .22/250 Remington. As is usually the case, IMR 4350 proved the best propellant selection for the test .270 rifle.

Accuracy is paramount for long-range varminting. At any range, a prairie dog is a mighty small target. Standing at 300 yards, that same prairie dog becomes a difficult challenge. Even a comparatively large 15-pound woodchuck is a hard-to-hit target when the ranges extend past 300 yards. So when selecting a rifle, it's best to place emphasis on accuracy and shootability rather than styling or aesthetics.

Rifles equipped with heavier barrels have reputations for providing better accuracy than those with skinny barrels. But heavier is not always better when it comes to field rifles. The hunter who walks a lot, or who simply would find an 11- or 12-pound rifle a burden in the field, may find a sporter his best choice.

Some shooters attempt to use benchrest-weight rifles (and barrel lengths) in the field. However, the extra weight of such rifles can take the fun out of long summer walks. Also, the short (20-inch) barrel length frequently found on some rifles deprives the hunter of some trajectory flatness, and the reduced velocity levels can create bullet expansion problems at very long ranges.

The Remington 700 series rifles enjoy a fine reputation for accuracy. The Ruger 77 and the Winchester Model 70 have very enviable reputations, and I can honestly say I have never heard an accuracy complaint about a Weatherby rifle when chambered for a varmint cartridge. And other brands of rifles will also provide fine accuracy. But do ask plenty of questions of friends, shooting acquaintances, or a trusted dealer, especially about the results with specific models in the caliber of your choice.

Above all, remember that accuracy is a subjective thing. A buddy who tells of 1/2-inch groups might well be referring only to five or six three-shot groups. Such accuracy, while admirable for most hunting, may not tell you what you want to know about long-range varminting potential.

A series of factors enter into long-range varmint rifle evaluation that may be ignored when evaluating a rifle or cartridge for a different purpose. Varmint rifles are often used in warm, even very hot weather. Thus, when used, the barrel may be 50° or even 70° F. warmer than a big-game rifle's barrel when it is used.

Varminters frequently have the opportunity for repeating a shot; sometimes a half-dozen rounds may be fired in rapid succession at pests.

Thus, five-shot groups fired as fast as the shooter can comfortably maintain his best accuracy should be the minimum testing standard. And it takes quite a few groups to establish a barrel's and load's average accuracy.

Barrels that are bedded with some amount of fore-end pressure against the barrel (as are most out-of-the-box factory rifles) are often quite accurate for three and sometimes five shots. But as the barrel heats and expands, the amount of fore-end pressure against the barrel can be the cause of shots stringing vertically on the target. For this reason, most varminters will free-float a barrel.

On rare occasions, such floating will actually enlarge groups. But, for the overwhelming majority of rifles, free-floating will not only prevent vertical stringing as the barrel heats, but will also improve a rifle's accuracy with a cold barrel.

Free-floating also offers the advantage of a constant point of impact. There will be no need to readjust sights as seasonal or extreme weather changes cause the stock to gain and lose moisture, thus varying its pressure on the barrel.

So, a good start in tuning with any rifle is to free-float the barrel. Then, after the barrel has been floated, it should be determined if the action is properly bedded. If problems are encountered, the action can be glass-bedded. However, I hasten to point out that, to the credit of many of today's manufacturing methods, I seldom find a need to rebed a quality rifle's action.

Accuracy is also dependent on the barrelled action, stock, mount, and scope functioning as one. This means not only that everything needs to be properly fitted, but everything needs to be tight—very tight. I've often solved others' shooting accuracy complaints by no more than the judicial application of the appropriate-size screwdrivers to the rifle system's various screws.

Be sure that you don't mistake a screw that bottoms in its hole as being tight. A screw that bottoms cannot apply the necessary tension on the parts it is supposed to be retaining.

When selecting a rifle, be sure to choose one

VARMINT CARTRIDGE SELECTOR

Caliber	Bullet Wt./ Type (grs.)	Powder Type	Charge (grs.)	Velocity[1] (fps)	Trajectory (ins.) at:[2] 200	300	400 yds.	Accuracy[3]	Recoil[4] (ft.-lbs.)
.224 Wby. Mag.	55 Factory	Wby. Factory		3638	+2.0	−2.4	−13.0	1.250″	4¼
″ ″ ″	50 Nosler HP	H335	30.0	3489	+1.6	−4.8	−18.3	0.700″	3½
.22/250 Rem.	55 Power-Lokt	Rem. Factory		3591	+1.8	−2.5	−13.5	0.988″	4¾
″ ″ ″	55 Speer Spit.	H380	39.5	3543	+1.8	−3.2	−14.7	0.668″	5
.220 Swift	55 Speer Spit.	IMR 4350	41.0	3444	+1.6	−4.0	−17.4	1.115″	5
.243 Win.	80 Power-Lokt	Rem. Factory		3254	+1.4	−4.1	−15.4	1.003″	8
″ ″	75 Sierra HP	IMR 4350	46.0	3300	+1.2	−5.4	−19.9	0.949″	7½
6mm Rem.	80 Power-Lokt	Rem. Factory		3225	+1.3	−4.2	−15.6	1.539″	8
″ ″	80 Speer Spit.	IMR 4350	46.0	3272	+1.6	−3.4	−14.0	1.109″	8
.240 Wby. Mag.	87 Factory	Wby. Factory		3410	+1.8	−3.1	−13.9	1.606″	10¾
″ ″ ″	75 Speer HP	IMR 4350	53.0	3575	+2.0	−2.8	−14.1	1.122″	9½
.25/06 Rem.	87 Power-Lokt	Rem. Factory		3393	+1.8	−3.1	−13.9	1.667″	10¾
″ ″	100 Speer Spit.	H4831	55.0	3200	+1.4	−4.1	−15.4	1.008″	11¾
.257 Wby. Mag.	87 Factory	Wby. Factory		3635	+2.0	−2.1	−11.3	1.5999″	14
″ ″ ″	100 Speer HP	IMR 4350	64.0	3255	+1.0	−5.9	−20.6	1.136″	14
.270 Win.	100 Soft Point	Rem. Factory		3149	+1.0	−5.5	−19.0	1.527″	10½
″	90 Sierra HP	IMR 4350	60.0	3532	+2.0	−2.9	−14.4	0.995″	12½
″	100 Speer Spit.	IMR 4350	59.0	3253	+1.4	−4.4	−16.3	0.852″	13

(1) Velocities recorded at 15 ft. from muzzle; (2) Trajectory of each at 100 yds. is +2.0 ins.; (3) for five, 5-shot groups at 100 yds.; (4) recoil calculated on a 9-lb. rifle weight

that can have its trigger pull adjusted to a crisp 2 to 3½ pounds. Today, because of potential liability problems, rifle manufacturers leave trigger pulls at weights hardly suitable for the finest accuracy. Indeed, I have encountered many a rifle whose trigger pull was notably heavier than the rifle's mass weight.

Only a well-qualified expert who understands the interaction of sear engagement, safety function, and lightweight pulls should attempt trigger adjustment. Most often this means employing the services of a good gunsmith who specializes in such work.

Varmint hunters have no need to carry a loaded rifle about, so lighter trigger pulls present no real safety hazard. But if you use a varmint rifle for other hunting purposes, in which a loaded chamber is normal, you may wish to modify the suggested trigger pull range upward.

Thus, rifle selection comes down to the bolt model or models that are available for your caliber selection. The chosen model should be selected on the basis of known accuracy, a good trigger that can be adjusted to a useful weight of pull, and its shootability.

By shootability I mean that the rifle should suit the individual's field-shooting style. Most rifles feel just fine when put to the shoulder in a gun shop. But usually, when in a belly-down position, there may be a specific model that is better suited to a shooter's style than others.

The easiest method to determine what suits best is to shoot different models under field conditions. This is not a practical approach for everyone, but a lot of questions can be asked of the most experienced long-range varminters you know. From a practical standpoint, those rifles that are popular with varminters are satisfactory.

I find the Weatherby Varmintmaster one of the most shootable varmint rifles I have ever used. And the now discontinued Remington 700 Classic is also a very shootable rifle for my style. But the best choice for you may be different.

This characteristic of shootability can, however, ultimately be answered only by the shooter who has gained sufficient experience with a number of different rifles to make possible an objective evaluation. What suits one shooter best may be one of the worst choices for another shooter.

I am appalled at the number of shooters who seem to believe that a 100-yard 1/2-inch group can be extrapolated into a 400-yard 2-inch group. It just isn't so. To maintain such a relationship would mean that your visual acuity, your ability to aim precisely, would have to be enhanced by a scope four times as powerful for the 400-yard shooting than was used for the 100-yard shooting. Aiming error simply is greater as the range increases when a scope of a given power is used at each range. Then there are winds and mirage that may vary over the long range.

Thus, it makes good sense to select a high-magnification scope for long-range varminting. However, there is a practical limit to how much

RIFLES USED FOR ACCURACY TESTING

Brand	Model	Caliber	Barrel Length (ins.)	Scope Brand & Power	Mounts	Barrel History[2] (rounds)
Weatherby	Varmintmaster	.224 Wby. Mag.	24	Leupold 12X	Redfield Jr.	1,400
Remington	700 Classic[1]	.22/250 Rem.	24	Leupold 12X	Redfield Jr.	2,800
Ruger	77V	.220 Swift	26	Zeiss 10X	Ruger rings	(unk.)
U.S. Repeating Arms	Win. Model 70 Varmint	.243 Win.	24	Zeiss 10X	Redfield Jr.	965
Remington	700 BDL	6mm Rem.	24[3]	Leupold 12X	Conetrol	(unk.)
Weatherby	Mark V	.240 Wby. Mag.	24	Leupold 12X	Redfield Jr.	315
Remington	700 ADL	.25/06 Rem.	24	Zeiss 10X	Redfield Jr.	(unk.)
Weatherby	Mark V	.257 Wby. Mag.	24	Leupold 12X	Redfield Jr.	255
Remington	700 ADL	.270 Win.	22	Zeiss 10X	Redfield Jr.	780

All rifles were tested with free-floating barrels and with trigger pulls adjusted between 2½ to 3 lbs. (1) Discontinued model; (2) Total of rounds fired through barrel to conclusion of testing including rounds fired previous to these specific tests; (3) Custom barrel

power can be employed in the field. It is one thing to use a 36X scope from a bench rest and quite another to attempt to use it in the field. Experience shows that only a rare shooter can employ 20X or more in the field.

Body tremors, breathing, and the movement caused by a less than rock-steady position are magnified in a scope. The more power employed, the more distracting the movement of the crosshairs on the target becomes. For most experienced shooters a scope of 15X or 16X is the maximum.

But there will be occasional shorter-range shots that might best be taken offhand or from a sitting position. Then 12X seems to be about the ideal compromise. Surely enough long-range varminting is successfully conducted with scopes of 8, 10, or 12X to establish the usefulness of these levels of magnification.

Too many riflemen fail to realize the difficulty in locating a target in the tiny field of view of a high-powered scope. The woodchuck so easily spotted with a binocular sometimes is more than just a tad difficult to find in a scope, especially for a beginner. A field of view of a few scant feet does not leave room for many reference marks; often just grass is all that can be seen.

Typically, 8 to 12X scopes might have 100-yard field of views of 15 to 9 feet, respectively. A typical 20, 24, or 36X scope might afford a field of view of only 3 or 4 feet. It's hard to assess the degree of potential difficulty in locating a target in such a small field of view, even at long ranges, until one actually must point his scope at a target surrounded by acres of lookalike terrain.

However, experience will, with time, make the task less and less a chore. Still, a 12X scope will make a 400-yard distant target appear as though it is but 33 yards away. That's enough magnifi-

cation for most shooting. And offhand shooting at shorter ranges can become very practical with such a scope.

Regardless of the power of the scope selected, keep in mind that the scope's relative brightness and clarity are extremely important. A magnified blur is not much of an aiming point. A bright, well-defined target is always a necessity. Get the best scope you can afford.

Because a long-range varmint rifle is used for extremely difficult shooting situations, the shooter's insight can help him select those rifles that have the best features to help make hits a reality.

For example, a rifle equipped with a no-slip rubber buttplate will usually be easier to shoot than one with a slippery plastic buttplate. A wide trigger is often an asset when compared to an unduly narrow one. A very high comb can make prone shooting a tad easier as can the mounting of the scope as low as possible. And the use of a sturdy and securely mounted bipod can add a great deal of hitability to your chosen rifle.

But even the simplest bipod is better than no bipod. And a realistic evaluation of the shooter's tolerance to recoil is a very wise step to undertake before selection of a rifle.

Don't forget the usefulness of a military rifle sling, especially if you do not use a bipod. A properly adjusted sling can make distant targets a lot easier to hit as compared to attempting the shot without a sling.

The selection of a suitable cartridge, rifle, scope, and maybe a bipod and/or sling are the first steps to effective long-range varminting. Careful consideration of your goals beforehand will help make your choices valid and lasting ones. Then you will need only to apply basic long-range marksmanship skills to begin enjoying long-range varmint shooting.

Synthetic Stocks: Technology Triumphs Over Tradition

Ross Seyfried

For hundreds of years, shooters have admired beautiful wooden gunstocks, so I'm hardly surprised that the question I hear most is: "Why on earth would I want one of those ugly plastic things?" This question follows the suggestion that someone try a "fiberglass" gunstock. The basic reason you should consider having a "plastic" stock on your pet rifle is that the stocks made of some synthetics perform well, better over the long haul than any wood. A wooden gunstock actually has only one virtue over fiberglass—wood is good to look at. There are degrees of "good to look at" that range from ho-hum, chair-leg wood, covered with too much shiny varnish like the ones found on most factory rifles, to stocks that can only be described as sensual,

magnificent pieces of God's art. When compared to overvarnished plain wood, some of the modern synthetic stocks with their matte finishes win the beauty contest in my eyes, but when it comes to gorgeous wood, with a proper finish, there is no contest. I am addicted to beautiful wood, plain and simple. I hoard the stuff like some idiots must hoard their drugs. There are stock blanks of every description—French, American, Bastogne, Claro, even some Myrtlewood—hidden in my attic and under my bed. They don't have a purpose. I don't know what action I will fit them to, but I possess them and that is good enough, just knowing that I have the raw material at hand

This article first appeared in Guns & Ammo

Synthetic stocks, once seen only on custom rifles, are now available on many factory rifles and also as drop-in accessories. Rifles shown here are (top) custom Model 70 in .338/06 with McMillan stock and Remington .30/06 with RAM-LINE stock.

Modern synthetic stocks can withstand real abuse. They are most impressive when subjected to temperature extremes. This resistance plus unusual strength make them a standout in the hunting fields. Right: Author Seyfried drove his vehicle over open action portion of RAM-LINE stock without harm to stock.

to create a masterpiece whenever I see fit. When I carry a beautifully stocked rifle or shotgun in the field, it is a source of pride. I look at it and enjoy its presence. I am more cautious about what touches my rifle because I don't want any scratches, and rain is a source of concern because it hurts the finish and accuracy. I also have rifles with synthetic stocks because I know nothing can hurt them.

When you get past the beauty contest, the synthetic stocks win hands down. They aren't affected by moisture or temperature. When we are talking about rifle stocks, the change in shape (warpage) caused by wetting or drying usually has a bad effect on accuracy or point of impact or both. Even if the wandering around of a wooden stock still lets the rifle shoot little groups, they don't count if the rifle won't put those little groups in the same place every day. When I shoot a working rifle, I am interested in "one-shot" groups. I am not nearly as impressed by a rifle that will shoot quarter-inch, 10-shot groups as I am by one that will shoot a "one-shot" group inside a 2-inch circle at a hundred yards—every day, week after week, cold, hot, wet or dry. Show me a rifle with a wooden stock that will do that and I will show you something as rare and desirable as a Boone and Crockett bull elk. They both occur, but don't bet your boots on being able to find one this weekend. On the other hand, I have found that a skilled gunmaker can turn out a fiberglass-stocked rifle that will meet those standards almost at will.

The other thing that makes synthetic stocks better than wood is that they don't break. Even the best straight-grained walnut is relatively fragile compared to the synthetic stocks, and the plastic stocks are usually lighter than their wooden counterparts even though they are several times stronger.

Ten years ago, a fiberglass-stocked rifle was hard to find. Now they are everywhere and you can get them three different ways. There are the custom stocks made from fiberglass blanks that are similar to semi-inletted wood blanks; factory rifles fitted with fiberglass stocks; and aftermarket synthetic stocks designed to let you drop your barreled action into them without any work.

To the best of my knowledge, Chet Brown made the first production fiberglass stocks. He is still making them, using fiberglass, Kevlar, and graphite either individually or in combination. Brown Precision and fiberglass stocks are almost synonymous. A vast majority of the synthetic stock technology is the result of Chet Brown's effort. You can still get anything from a rough blank to a finely finished custom rifle from Brown.

You can now buy Remington rifles stocked with state-of-the-art fiberglass/Kevlar/graphite stocks made by Brown Precision. In addition to the Remingtons, I have been using out-of-the-box fiberglass-stocked rifles from Weatherby,

Sako, and Winchester. These three companies all have their fiberglass stocks made by another of the originators of the art, Gale McMillan of McMillan & Co., Phoenix, Arizona.

A very interesting way to get a rifle with a synthetic stock is to buy one of the aftermarket add-on synthetic stocks. They are available from several makers, including McMillan, but the Millett and the new RAM-LINE stocks are the only ones I have personally tested. The idea behind these stocks is that you can go to the store and buy a synthetic stock that will fit the "old Betsy" you have in the gunrack. This idea is new, and there is a wide range in the price and quality of the products. But being able to slap the barreled action of your favorite wood-stocked rifle into a synthetic stock at will and have it perform under tough conditions will be a growing trend in the future.

When I started to prepare this article, I realized that my personal battery of synthetic-stocked rifles was quite limited and that I hadn't had first-hand experience with fiberglass-stocked rifles except for custom ones. My first fiberglass rifle had a Brown stock blank that I fitted to a Shilen barreled action chambered for .22/250 Improved. Contrary to the belief that all fiberglass rifles are lightweights, this rifle weighs 13½ pounds. When I fitted the 26-inch, No. 7 contour barrel into the Brown varmint-pattern stock, the rifle was very muzzle-heavy. I happen to believe, like the British gunmakers, that no matter what weight a gun is, it should balance. To get the rifle to balance on the front guard screw, I added a mix of No. 7½ lead shot and epoxy to the butt until the rifle's balance was right. This monster wasn't intended for running coyotes down; instead, it was designed to shoot them from where I stood. The final rifle, painted with bronze Du Pont Imron paint, topped with a 4X–12X Redfield

Brown Precision .416 Hoffman Professional Hunter with nickel-plated finish, Leupold 1.5X-5X scope, and integral muzzle brake, but minus hinged metal floorplate, tips scale at easy-to-carry 7.8 pounds.

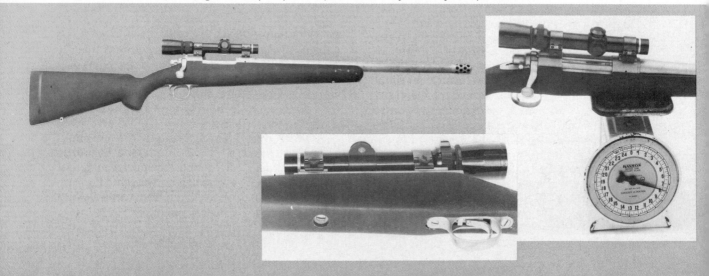

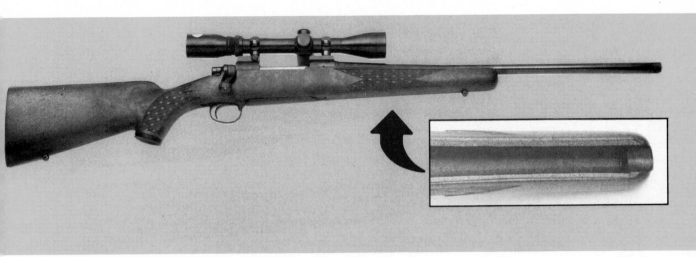

Remington 700 .30/06 set in RAM-LINE synthetic stock and topped with Burris 3X-9X scope makes a very tough hunting combo. Forearm of stock required minor bedding work for perfect fit. Barrel is not free-floated; there is pressure at the tip.

scope, and driving 55-grain Noslers at 4,000 fps into quarter-inch groups, is a fur-coat machine. It is over 10 years old, has flattened enough fur to make a long stretch of Fifth Avenue happy, *and* has never changed zero.

Another rifle that I am very familiar with is a Remington Model 700 .270 Winchester with its barrel turned down to a minimum contour and fitted into a Brown Kevlar stock. Topped with a 2X–7X Leupold scope, it weighs just over 6 pounds. It was made for a friend to hunt Argali sheep in Mongolia, where the license costs more than a small truckload of good rifles, not to mention the dreams of a lifetime wrapped up in the shot at such a great sheep. I can tell you that making this rifle made me just a little nervous.

When it was finished, I worked up a load with 130-grain Nosler Partition bullets that would shoot 1-inch groups until you were tired of watching. I then carried, banged, froze, and heated the rifle for weeks to be sure the point of impact didn't shift. When it came time to shoot the great sheep the range was over 200 yards and the bullet had to pass through the curl of the ram's 66-inch horns to find his shoulder. He was one of the largest Argali ever taken. If I had the chance to make the same hunt, my stock wouldn't be made of wood!

I came to know both of these rifles very well, but I wanted to learn a lot more about production fiberglass rifles, so I ordered one of everything I could get my hands on. This was a very rewarding test in that all shooting was done with factory ammunition and untuned, out-of-the-box rifles. I added a RAM-LINE stock to a Remington 700 BDL .30/06 and a Millett stock to a Ruger Model

77 chambered for .270 Winchester. The rest were factory—that is, McMillan, synthetic-stocked rifles from Weatherby, Winchester, (USRAC), and Sako. I used ammunition from every major manufacturer: PMC, Winchester, Remington, Hornady, Federal, and Weatherby (Norma). Throughout all of my shooting the largest 100-yard groups I shot were 2 inches, while most of the groups were under 1½ inches. In an overview, our arms and ammunition manufacturers are doing one hell of a good job.

The Millett stocks are designed for an instant fit on your factory Remington, Winchester, or Ruger rifle. These stocks come with molded-in checkering, a crackle finish, and in colors ranging from black to orange. Millett stocks also have a quickly adjustable nylon sling that comes with the stock. Instead of using conventional sling-swivel studs, the Millett has small rods molded into the forearm and buttstock that hold the sling in place. You disassemble the sling to remove it. I would prefer conventional swivels, but that is just a personal matter of choice. My test sample from Millett was for a Ruger M77. When I put the Ruger in the stock, I found the inletting was too tight in some places. I phoned Millett to discuss the problem and I found that my stock was a rush job to help me make a magazine deadline. My stock had bypassed final inspection to get it to me in a hurry. I worked with my inletting scrapers and soon had a fine fit for the Ruger, and had the rifle shooting as well as it would in its factory stock. Groups under an inch were common. (Millett told me that occasionally tolerances in both rifles and stocks will produce a combination of barreled action and stock that

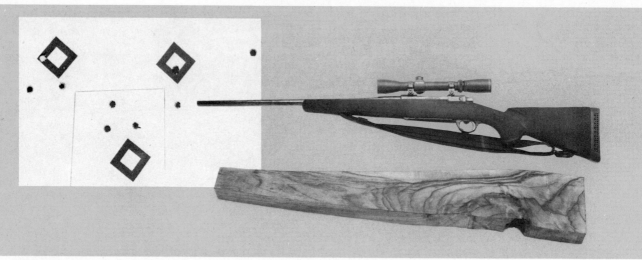

Beautifully figured wood blank shown here can be turned into work of art, but Ruger 77. 270 Winchester in Millett stock with Pentax scope is virtually weatherproof. 100-yard groups were fired with Ruger M77 .270 in factory wood stock (upper left), same rifle in Millett stock (upper right) and in Millett stock after bedding work.

won't fit properly. If you encounter difficulty fitting a rifle into your Millett stock, they want you to phone them to discuss the trouble before you try to alter your stock. Millett wants you to have a stock that is an instant fit on your rifle, and they will take care of any problems you might have.)

I also had a chance to try the very new RAM-LINE "newfangled" stock. This model fit a Remington 700 BDL .30/06. The RAM-LINE is a radical departure from ordinary synthetic stocks. Most all other synthetic stocks are formed like a boat hull, with layers of cloth (fiberglass, Kevlar, or graphite) layered in a mold and "glued" together with some sort of epoxy resin. The RAM-LINE stock is injection-molded. To make these stocks, a mixture or aerospace polymers and glass fibers are forced into a mold in liquid state, using 600° F. and 20,000 psi pressure to get the job done. The results are extremely precise and so strong I can hardly believe it.

RAM-LINE told me I could drive a car over their stock without damaging it. I took them at their word and put their stock (without the barreled action) on the frozen ground and drove the front tire of my Blazer over it. I didn't drive over a solid part, but the hollow portion where the magazine box goes. In the process I wanted to take a picture and touched the Blazer's brake just as the tire was on top of the stock. When I touched the brake the tire slid a little on the stock and caused it to roll under the tire. The stock turned 90 degrees, right over the fragile edges of the inletting—and suffered only a few scratches from the rocks. My Remington fit the stock

nicely, but there were a few places that a fine file and some sandpaper would help. My barrel channel was a little tight; gentle sanding on the sides of the inside of the stock made a perfect fit. There were also a few cosmetic places where the mold left slight seams, particularly on top of the comb, where a few strokes with a file make the stock a lot nicer. On the right side of the stock, just behind the receiver ring, there is excess material that can be removed to make the ejection port on the stock a perfect fit.

The RAM-LINE stock is a sort of "pearlized" black color and has checkering panels molded into it. The stock is fitted with studs for quick-detachable sling swivels and can be shortened over 2 inches to accommodate shooters with shorter arms or lengthened by the addition of recoil pads. Initially, stocks are available for Remington 700 BDL, Ruger Model 77, and Winchester Model 70s. As they come out of the box, the bedding should be as precise as factory wooden stocks, with variations in the rifles themselves making the kind of ultra-precise bedding found in custom rifles impossible. This looks like a winner and you can have a wooden stock plus an indestructible, weather-impervious, synthetic stock for much less than two rifles would cost.

The Weatherby Fibermark rifle I tested, with its McMillan-produced stock, is going to be hard to send back now that the testing is over. First it came chambered for one of my favorite cartridges, the .340 Weatherby. This sleek black beauty looks like a business rifle, unlike the usual Weatherbys with their all-too-bright polish and blue metal and shiny stocks. The Fibermark has

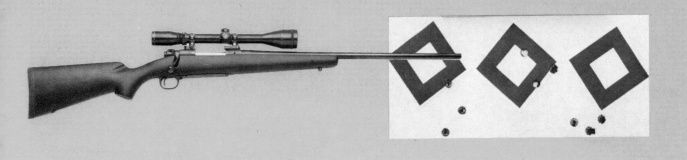

Winchester M70 with factory McMillan stock and Burris 4X-12X scope produced these tight 100-yard benchrest groups fired at 2-inch squares with (from left) Winchester, Remington, and Federal Premium .30/06 ammo.

a matte, vapor-blasted finish on the steel and a black crackle finish on the stock. Topped with Buehler mounts and a 3X–9X Weatherby scope, this brute would hammer 250-grain factory ammunition into groups that were regularly under an inch. Using handloads with premium big-game soft-noses and Barnes Super Solids, this rifle is fit for almost any hunting task on earth. The Fibermark shows some hand-bedding in the inletting, touches that make it shoot and cost much like a custom rifle. I like it, and am going to try to sell something so I can make this rifle a permanent addition to my gunrack.

The Model 70 Winchester made by United States Repeating Arms is their Model 70 XTR barreled action fitted to a McMillan fiberglass stock. My test sample was a .30/06 that was both accurate and very consistent about where it put its groups. I had the feeling that the rifle wanted be superaccurate, but needed a little work on the barrel bedding in the fore-end. It tended to string its shots a little left and right while holding less than a ½-inch dispersion in elevation. I am sure that with an individual bedding job and selected handloads this rifle will shoot ½-minute groups! If you like Winchesters and need a working rifle, this would be hard to beat. In fact, this is one of those cases where I think the plain black fiberglass rifle looks better than its wood-stocked counterpart with a thick coat of varnish. Of course, we can continue to hope that someday USRAC will give us the pre-'64 Model 70 back and fit it in the new stock. That would be special.

The Sako Finnbear .30/06, called the Fiberclass,

is an interesting rifle with a stock also produced by Gale McMillan. It was different from all the other production rifles and stocks because it came with its barrel free-floated instead of having pressure applied to the barrel by the fore-end. In my opinion, this is how most rifles should be bedded, and certainly all rifles with synthetic stocks should have this free-float barrel bedding. For one thing, fiberglass and bedding master Chet Brown tells me that fiberglass stocks will take a "set" over time. That is, the stock will tend to bend with the barrel pressure and stay bent. The result is that a stock initially fitted with 7 pounds of upward pressure on the barrel will soon lose most of that tension, and in the process change the point of the rifle's impact. If the barrel doesn't touch the for-end at all, this change isn't a consideration. This is the way the Sako Fiberclass is bedded. There is daylight between the stock and barrel for the entire barrel length. I am sure that this is why the Sako exhibited the most consistent point-of-impact performance of all the production rifles I tested. I used this rifle in conjunction with Hornady 168-grain hollow-point boattail match ammunition for the most important test in this entire article. I sighted the Sako to hit dead center at 100 yards with the Hornady ammunition. The rifle was topped with a 4X Swarovski scope that is one of the most reliable scopes I have ever used. The Sako would almost always shoot three-shot groups under an inch with the Hornady match ammo. When I finished fine-tuning the scope, I left the rifle with its metal parts

Weatherby .340 Fibermark (below) in McMillan stock with Buehler mounts and Weatherby 3X-9X scope proved accurate, due in part to care taken in bedding recoil lug/magazine box.

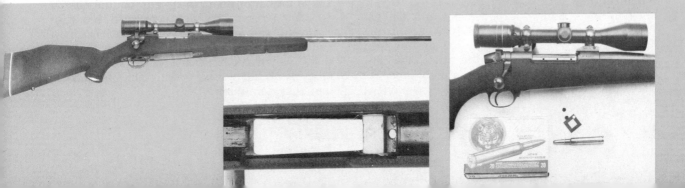

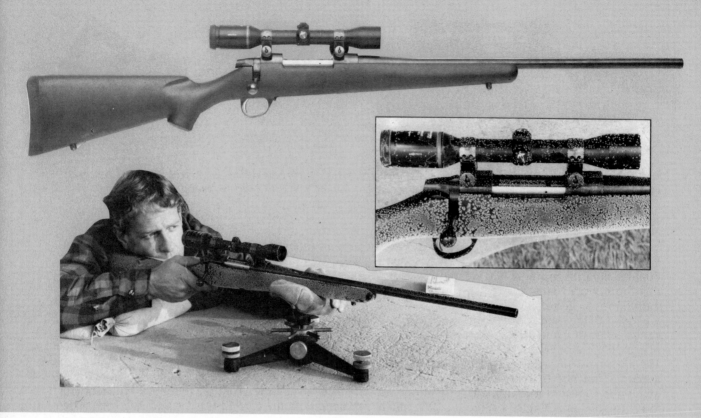

Sako Fiberclass .30/06 with McMillan stock and Swarovski 4X scope withstood author's freezing-cold/burning-heat torture test with flying colors. Ability to maintain zero is synthetic stock's strong suit.

well oiled (not the bore) on my shooting bench. It was December in Colorado and over the three days and nights that the Sako lived outside the temperature ranged from below zero to 50° F above. In the mornings the rifle was coated with heavy frost, and the rising sun would then quickly warm the black rifle so that water ran into the inletting. On the third morning, just at sunrise, with the temperature 10°, I fired one shot out of the frost-covered rifle. The target was a 2-inch black diamond with a 1-inch white diamond center at 100 yards. The bullet hit the center white diamond—impressive. Next I brought the rifle inside, put it under my woodstove and let it heat up until it was almost unpleasant to touch—just like an October day in the Zambezi valley or August in Arizona. I stepped out the door to my shooting bench and fired the hot rifle. That shot hit the white center too. Now that is a rifle you can trust. It is just the kind of performance I have come to expect from all rifles with synthetic stocks, floated barrels, and precise action bedding.

To complement my experience with other custom fiberglass-stocked rifles, I borrowed a rifle from Brown Precision chambered for the .416 Hoffman cartridge (.416 bore using a full-length .375 H&H case with most of the body taper removed). I would call this rifle an ultra-modern

.416 Rigby. It pushes the 400-grain bullets at 2,400 fps and puts them in tiny groups. The rifle is Brown's Professional Hunter model. The fiberglass stock has classic lines and is designed for minimum weight and bulk. The .416 elephant-stopper weighs less than 8 pounds, topped with a Leupold 1.5X–5X scope. Every square inch of metal is nickel-plated with a dull matte finish. This rifle is a no-compromise machine that you can stake your life on. It comes as close to being "dishwasher safe" as any rifle in existence, with the plated steel being more or less protected from the elements. The only thing better would be if we could get Ruger to make us an all-stainless-steel rifle for Brown Precision to put in a synthetic stock. Right now Brown is advertising a Model 70 Winchester in .270, .30/06, or 7mm Magnum, precision-bedded in one of their fiberglass/Kevlar stocks at a bargain price for a custom-fitted rifle.

I haven't mentioned all of the sources of synthetic stocks here, but I think I have covered most. I really like the looks of wooden stocks a whole lot better than plastic gun handles. But I do have a mental conflict, because mixed with my love of fine wooden gunstocks there is a hard line of practicality. When my life might be at stake, or when I hunt coyotes to pay the bills, my rifles usually have plastic stocks—because they just plain work.

The Wayward Wind

Jim Carmichel

There's no way I could have missed," the hunter said, rising slowly and unsteadily from a prone shooting position. "The crosshairs were dead still on the buck's shoulder, and I didn't flinch," he added.

For a moment, he stood there, groping for an explanation and desperate for an alibi.

"It was the rifle," he finally told his partner. "The rifle let me down. The scope isn't zeroed right. Yeah, that's go to be it," he insisted. No longer was the rifle his friend; now he regarded it as a once-faithful ally that had become a cunning deserter.

"The sights were okay when you checked them yesterday," the older hunter said. "Let's go—we've got a lot of walking to do before we catch up with that buck. Next time, it's my shot.

Still in shock, the hunter retrieved the treacherous rifle, pulled his hat low, and squinted into the Wyoming wind.

The wind! If the hunter had considered the air moving around him, he would have known the cause of his miss and he could have offered one of the best excuses a rifleman can make for missing.

Target shooters routinely blame the wind for poor scores. In fact, the ability to "dope" wind is the primary difference between champions and also-rans in many rifle tournaments. Yet, the effect of wind on a bullet's journey is largely ignored by hunters. We all realize, of course, that wind blows a bullet off course to some degree, but precisely or even approximately how much remains a mystery for many sportsmen. I've known hunters to confer in great detail about the distance to the target, determine the bullet's trajectory impact point to within fractions of an inch, and then only give scant consideration to the wind. Many times, I've heard guides giving their hunters detailed advice on range and bullet

This article first appeared in Outdoor Life

For long cross-country shots like this, knowledge of wind's effect on bullet's flight can mean the difference between solid hit and complete miss.

drop and then adding, "And allow a little for the wind."

The fact is that allowing for the wind is at least as vital to good bullet placement as accurate range estimation. If you doubt this for a moment, consider that at 300 yards a crossing 20-mile-per-hour wind will carry a 150-grain bullet from your beloved .300 Winchester Magnum a full 18 inches off course. At that same distance, the bullet will drop only some 6 inches below where you aim, if the rifle is zeroed for 200 yards. Ask yourself: Which should be cause for more worry when you launch a bullet across the wind-swept plains and mountains, the trajectory or the wind? The answer is obvious. Clearly, the wind is a formidable force to deal with, and it causes a lot more misses than shooters realize.

The thing that makes dealing with the wind a guessing game is that it is constantly changing. It's like trying to bargain with an Arab horse trader who never shows you the same animal twice. By the time you figure out what to do, it's a whole new game. To make a bad situation even worse, the fickle wind won't even do you the courtesy of blowing at an even rate at all points between you and the target. There may be a 20-mph wind at your shooting position, but it may be blowing at only 15 mph 100 yards farther out, and there may be a dead calm at the target, or the whole situation might be reversed. Long-range target shooters who try to hit a hat-size target at 1,000 yards often have to outguess winds that blow in two or more directions at the same time over the long distance the bullet will travel. All this is especially depressing to shooting buffs who dote on the precision of ballistic arithmetic and who carefully memorize trajectory figures.

HOW TO BEAT THE WIND

Happily, there are ways to beat the wind's effect, and if you figure out a way to cheat, that's all the better. As you study our wind-drift tables, you'll notice that certain calibers are much less affected by wind than others. You'll also discover that bullet velocity, weight, and even shape are important factors in winning the wind game. This is dramatically revealed by comparing the wind's effect on 180-grain pointed and round-nose bullets fired from a .30/06. Though both bullets leave the muzzle at 2,700 feet per second, the less-streamlined round-nose bullet slows more rapidly than the spitzer and is therefore affected more by the wind. At 300 yards, the pointed bullet is deflected 8.8 inches by a crossing 10-mph breeze, while the round-nose version

is carried off course nearly twice as far (14.8 inches).

This tells us that the same calibers and bullets that have flatter trajectories and good retained energy are also less affected by wind. A high-velocity cartridge firing a streamlined bullet can be your ace in the hole when betting against the wind.

Wind drift is among the most-complex ballistic subjects. Even though computers such as the one used at OUTDOOR LIFE's Briarbank Ballistic Institute to generate the tables can precisely calculate wind drift, there is some disagreement among ballistic experts about how the wind's force is exerted. Gun writers have occasionally likened the path of a bullet through wind to the movement of a boat crossing a river. This is not a fair comparison because of significant differences between boats and bullets. A spinning bullet in flight is a powerful gyroscope, and like any gyroscope, when it is pushed, it pushes back. Added to this is the force of the bullet's momentum. As Newton taught us, a body in motion likes to remain in motion and doesn't like to change directions. That's why a bullet reacts only marginally to the wind's strength. In order to understand this more clearly, let's use a specific example.

If we convert miles per hour to inches per second, our arithmetic will show that a 10-mph breeze is also blowing at 176 inches per second. If a bullet simply drifted in air currents like a leaf in a river, it would be carried off course 176 inches for every second of flight (and proportionately for fractions of a second) by a 10-mph crosswind. If so, the wind-drift situation would make it almost impossible to hit a target in the slightest wind. A 180-grain bullet from a .30/06 would be blown off course more than 20 inches in the .116 second it takes to travel 100 yards in a 10-mph wind. In the real world, the bullet is driven off course only about 1/20 of that amount (about one inch), which amply illustrates the bullet's built-in momentum and gyroscopic stability. These two unhailed blessings make it reasonably possible for us to hit a target when the wind blows.

RIMFIRE WIND DRIFT

Of course, there are exceptions to almost all rules, and the rules of ballistics have their own peculiar variations. When you take a close look at the .22 Rimfire wind-drift tables, you'll probably be surprised to find that there are instances in which a slow-moving (1,150 fps) bullet is less disturbed by wind than an identical .22 Rimfire bullet traveling somewhat faster (1,255 fps). This is because of some peculiar things that happen when a bullet travels at or close to the speed of sound, which happens to be the velocity neighborhood in which most rimfire cartridges live.

The reasons for this phenomenon are too involved to explain here, and results are more important to shooters. Before this blip on the ballistic screen was known and understood, it was common for small-bore (.22 rimfire) target shooters to use high-velocity ammo in the belief that this would be an advantage in windy conditions. It didn't work out that way. That's why the best rimfire match ammo is loaded to surprisingly low velocity levels. If you take your .22 rimfire out plinking on a windy day, keep in mind that you'll do better with standard-velocity loads. But keep in mind that this ballistic aberration occurs only in the transonic velocity range. When the velocity is upward of 2,000 fps, the hard-and-fast rule is that fast, streamlined bullets perform better in wind.

Though we can cheat wind to some degree by using hot calibers and fancy bullets, we still have to apply some solid know-how to get a bullet dead on target. Though there is no substitute for experience when it comes to judging or "doping" the effect of wind, there are a few tricks that will help.

THE "CHICKEN METHOD"

Though applying "Kentucky windage" by holding into the wind is a time-honored way of correcting for bullet drift, the real trick is knowing just how far to hold off the target to compensate for the wind's effect. This can be almost impossible, even for very experienced shooters, in a tense hunting situation when there isn't time to contemplate the range and study the wind. That's when I apply my "chicken method." It's not exactly foolproof, but at times, it works better than anything else, especially when the wind velocity and the target distance are so uncertain that it would be a real gamble to hold completely off the target.

In these cases, I simply hold the scope's vertical crosshair on the windward edge of the target area. In this way, I have the entire width of the target area as a margin for error. To describe my chicken technique in action, let's say that a nice pronghorn is facing me head-on at a distance of somewhere between 200 and 300 yards. The variable wind is blowing from right to left, and its velocity is too unsteady to estimate with any certainty. I put the vertical wire on the windward edge of the animal's outline. Even without knowing how much the wind will drift the bullet, I do know which way the bullet will be blown and I can be reasonably certain that it will impact some-

where in the vital area of the chest. Because the vital area of a pronghorn's chest is about 10 inches wide when seen from the front, I have a comfortable margin for error with a high-velocity cartridge. If the animal is standing sideways or quartering, I simply hold on the windward edge of the vital area. Shooting in the wind is no game for the timid. More shots are missed by underestimating wind velocity than by overestimating. When you study the wind tables, you'll see why.

My chicken method also works pretty well in varmint shooting. Holding on the windward edge of a fat eastern woodchuck usually solves the wind problem. Potting skinny prairie dogs in a windswept dog town requires a bit more finesse. If holding on the windward edge of the target doesn't work (and it won't in a stiff crosswind), I then hold off in multiples of body width. Once I connect, I have a reference to use for later shots.

GUSTS, LULLS, AND MIRAGE

I always follow two rules when shooting in the wind. I never shoot during a strong gust, and I never shoot during a lull. The first rule is self-evident, but the second takes some explaining because it seems sensible to fire when there is a lull in the wind. The risk of shooting during a lull is that at such times, the wind may suddenly come on strong from a new direction. A sudden lull often indicates that a change in direction is imminent. That's why savvy target shooters never fire during a lull. Instead, they try to evaluate and adjust for steady and stable wind speed and direction and fire only when those particular conditions exist.

Savvy shooters also learn to read mirage, the squiggly heat waves that are usually visible through a rifle scope or spotting scope. These rising heat waves, which are actually atmospheric distortions caused by rapidly changing air density, tell you a lot about what the wind is doing. As my pal Herb Hollister says, "Mirage is wind you see."

The rising heat waves dance in the wind like a stand of wheat, responding instantly to every change of force and shift of direction. The great thing about reading mirage is that doing so tells you what the wind is doing close to the target, which is more valuable than knowing only the wind conditions near the firing point. Long-range target shooters, especially those who win, routinely focus their spotting scopes at various distances between themselves and the target so they will know what the wind is doing all along the bullet's route.

When shooting at 50 yards or beyond, I focus

my spotting scope short of the target so that I can just barely see the bullet holes. In this way, I also get a better look at the mirage. Likewise, when I'm varmint shooting, I focus my high-magnification riflescope at an intermediate distance (usually at 200 yards when I anticipate 300-yard shots) so as to keep an eye on the mirage. This induces some parallax, of course, but not enough to worry about, and the additional wind dope more than compensates.

Though there are many refinements to reading mirage, the hunter can benefit by remembering only a couple of things. The first of these is that if the mirage appears to be running in horizontal lines, the wind velocity is 10 mph or faster. Lesser wind speeds can be determined quite accurately simply by observing the angle of the mirage. Of course, if the mirage is boiling straight up, there is no wind, but beware. That could signal an imminent shift in wind direction. Wind in your face or at your back has very little effect on a bullet's flight, so in most hunting situations, it can be ignored.

I've tried to learn wind "guesstimating" by holding my arm out of a car window at various speeds, but it's easy to be fooled. At wind speeds more than 20 mph, it is difficult to judge within 10 mph. All things considered, my chicken technique works best in many situations.

THE TABLES

The accompanying wind tables show bullet drift caused by a 10-mph crosswind for 122 popular rifle, rimfire, and handgun loads. The drift caused by winds of other velocities is easily calculated by multiplying or dividing the values given in the tables. For example, to determine the effect of a 15-mph wind, simply multiply the listed value by 1.5

During the planning of this article, it was suggested that the tables list wind effect only out to 300 yards. The reasoning was that to list wind drift at longer ranges could encourage reckless and unsportsmanlike shooting at excessive range. I agreed with that wholeheartedly until I saw the tremendous amount of drift that does occur at the long ranges. I now think that, if anything, our wind tables reveal the hopelessness of attempting long shots when the wild wind blows out yonder.

DEFLECTION CAUSED BY A 10-MPH CROSSWIND

Caliber	Bullet Weight And Type* (Grains)	Muzzle Velocity (Feet Per Second)	Wind Drift (Inches)				
			100 Yards	200 Yards	300 Yards	400 Yards	500 Yards
Centrefire Cartridges							
.17 Remington	25 HP	4040	1.4	6.3	15.7	31.5	56.1
.22 Hornet	45 PSP	2690	2.9	13.5	34.9	66.9	106.6
.222 Remington	50 PSP	3140	1.7	7.3	18.3	36.4	63.1
.222 Remington	55 PSP	3020	1.5	6.5	16.1	31.6	54.2
.222 Remington Mag.	55 PSP	3240	1.4	5.9	14.5	28.4	48.9
.223 Remington	55 PSP	3240	1.4	5.9	14.5	28.4	48.9
.22/250 Remington	55 PSP	3680	1.1	4.9	11.8	22.7	38.8
.30/06 Accelerator	55 PSP	4080	1.0	4.4	10.7	20.6	35.1
.30/30 Accelerator	55 PSP	3400	1.3	5.6	13.6	26.5	45.6
.243 Winchester	80 PSP	3350	1.0	4.3	10.4	19.8	33.3
.243 Winchester	100 PSP	2960	1.0	4.1	9.7	18.2	30.0
6mm Remington	80 PSP	3470	1.0	4.2	9.9	18.9	31.7
6mm Remington	100 PSP	3100	0.8	3.3	7.9	14.7	24.2
.25/35 Winchester	117 PSP	2230	2.2	9.7	23.6	44.7	72.1
.250 Savage	100 PSP	2820	1.1	4.9	11.7	22.2	37.2
.257 Roberts	117 RN	2650	1.5	6.5	15.8	30.5	51.4
.25/06 Remington	120 PSP	2990	0.8	3.5	8.1	15.2	25.0
6.5mm Remington Mag.	120	3210	0.8	3.5	8.4	15.7	25.9
.264 Winchester Mag.	100 PSP	3320	1.0	4.4	10.6	20.2	34.0
.264 Winchester Mag.	140 PSP	3030	0.8	3.2	7.4	13.8	22.7
.270 Winchester	130 PSP	3060	0.9	3.6	8.5	16.0	26.5

HP, Hollow Point; PSP, Pointed Soft Point; SP, Soft Point; RN, Round Nose; BP, Bronze Point; PP, Power-Point; PP/FP, Power-Point/Flat Point; ST, Silver Tip; FP, Flat Point; JHP, Jacketed Hollow Point; NP, Nosler Partition; FMC, Full Metal Case; STHP, Silver Tip Hollow Point; JSP, Jacketed Soft Point; SJHP, Semijacketed Hollow Point

Caliber	Bullet Weight And Type* (Grains)	Muzzle Velocity (Feet Per Second)	Wind Drift (Inches)				
			100 Yards	200 Yards	300 Yards	400 Yards	500 Yards
Centrefire Cartridges (cont.)							
.270 Winchester	130 BP	3060	0.8	3.2	7.6	14.2	23.3
.270 Winchester	150 RN	2850	1.2	5.3	12.9	24.6	41.5
7mm/08 Remington	140 PSP	2860	0.8	3.4	7.9	14.8	24.3
.284 Winchester	125 PP	3140	0.9	3.8	9.0	16.9	28.1
.284 Winchester	150 PP	2860	0.9	3.9	9.2	17.2	28.4
.280 Remington	150 PSP	2970	0.9	3.7	8.7	16.2	26.7
.280 Remington	165 PSP	2820	1.1	4.8	11.4	21.7	36.3
7mm Remington Mag.	150 PSP	3110	0.8	3.4	8.1	15.1	25.0
7mm Remington Mag.	175 PSP	2860	0.7	3.1	7.2	13.3	21.7
.30 Carbine	110 SP	1990	3.4	15.0	35.5	63.2	96.7
.30/30 Winchester	150 FP	2390	2.2	9.8	24.3	46.7	75.9
.30/30 Winchester	170 FP	2200	1.9	8.0	19.4	36.7	59.9
.300 Savage	180 RN	2350	1.7	7.5	18.2	34.8	57.5
.300 Savage	180 PSP	2350	1.1	4.6	10.9	20.3	33.3
.307 Winchester	150 PP/FP	2760	1.7	7.6	18.8	36.9	62.6
.307 Winchester	180 PP/FP	2510	1.5	6.6	16.1	30.9	51.7
.308 Winchester	150 PSP	2820	1.0	4.4	10.4	19.7	32.7
.308 Winchester	180 RN	2620	1.5	6.4	15.5	29.7	50.0
.308 Winchester	180 PSP	2620	0.9	3.9	9.2	17.2	28.3
.30/06 Springfield	150 PSP	2910	1.0	4.2	9.9	18.8	31.2
.30/06 Springfield	150 BP	2910	0.9	3.6	8.4	15.7	25.8
.30/06 Springfield	165 PSP	2800	1.0	4.1	9.6	18.1	29.9
.30/06 Springfield	180 PSP	2700	0.9	3.7	8.8	16.5	27.1
.30/06 Springfield	180 RN	2700	1.4	6.1	14.8	28.4	47.9
.30/06 Springfield	180 BP	2700	0.9	3.7	8.8	16.5	27.1
.30/06 Springfield	220 RN	2410	1.4	5.9	14.3	27.1	44.9
.300 Holland & Holland	180 PSP	2880	0.8	3.4	8.0	15.0	24.6
.300 Winchester Mag.	150 PSP	3290	0.9	3.8	9.0	16.9	28.1
.300 Winchester Mag.	180 PSP	2960	0.7	2.8	6.7	12.3	20.0
.303 Savage	190 ST	1890	2.3	10.0	23.6	43.1	67.5
.303 British	180 RN	2460	1.6	7.1	17.2	33.0	55.0
.32/20 Winchester	100 RN	1210	4.3	15.6	32.6	55.2	83.2
.32 Winchester Special	170 FP	2250	1.9	8.4	20.4	38.7	63.2
8mm Mauser	170 SP	2360	2.2	9.5	23.0	44.5	72.4
8mm Remington Mag.	185 PSP	3080	1.0	4.1	9.7	18.3	30.5
8mm Remington Mag.	220 PSP	2830	0.9	3.7	8.7	16.3	26.8
.35 Remington	150 PSP	2300	2.5	11.0	27.2	51.5	82.3
.35 Remington	200 RN	2080	2.7	12.0	29.1	53.5	83.6
.338 Winchester Mag.	200 PSP	2960	1.0	4.2	9.9	18.7	31.2
.338 Winchester Mag.	225 SP	2780	0.8	3.1	7.3	13.6	22.1
.351 Winchester	180 RN	1850	2.6	11.1	26.2	47.3	73.2
.356 Winchester	200 FP	2460	1.7	7.3	17.9	34.3	57.2
.356 Winchester	250 FP	2160	1.5	6.6	15.6	29.4	48.1
.358 Winchester	200 RN	2490	1.5	6.5	15.7	30.0	50.0
.358 Winchester	250	2250	1.4	5.9	14.0	26.3	43.2
.375 Winchester	200 FP	2200	2.2	9.8	23.81	45.0	72.3
.375 Winchester	250 PP	1900	2.1	8.8	20.8	38.3	60.7
.375 H&H Mag.	270 SP	2690	1.1	4.5	10.7	20.2	33.6
.375 H&H Mag.	300 RN	2530	1.7	7.2	17.6	34.0	56.9
.38/55 Winchester	255 SP	1320	2.2	8.6	18.8	32.1	48.4
.44 Remington Mag. (rifle)	240 JHP	1760	4.0	16.9	38.0	65.4	98.3
.444 Marlin	240 JHP	2350	3.1	14.2	35.3	65.4	102.3
.444 Marlin	265 FP	2120	2.7	11.6	28.3	52.4	82.4
.45/70 Government	300 HP	1880	2.0	8.5	20.2	37.1	58.7
.45/70 Government	405 RN	1330	2.8	10.8	23.2	39.3	58.8

*HP, Hollow Point; PSP, Pointed Soft Point; SP, Soft Point; RN, Round Nose; BP, Bronze Point; PP, Power-Point; PP/FP, Power-Point/Flat Point; ST, Silver Tip; FP, Flat Point; JHP, Jacketed Hollow Point; NP, Nosler Partition; FMC, Full Metal Case; STHP, Silver Tip Hollow Point; JSP, Jacketed Soft Point; SJHP, Semijacketed Hollow Point.

DEFLECTION CAUSED BY A 10-MPH CROSSWIND (continued)

Caliber	Bullet Weight And Type* (Grains)	Muzzle Velocity (Feet Per Second)	Wind Drift (Inches)				
			100 Yards	200 Yards	300 Yards	400 Yards	500 Yards
Centrefire Cartridges (cont.)							
.458 Winchester Mag.	500 RN	2040	1.5	6.3	15.0	27.9	45.3
.458 Winchester Mag.	510 RN	2040	1.9	8.2	19.6	36.6	58.9
Weatherby Cartridges							
.224 Weatherby	55 PSP	3650	1.0	4.1	9.8	18.6	31.3
.240 Weatherby	87 PSP	3500	0.8	3.3	7.8	14.6	24.1
.240 Weatherby	100 PSP	3395	0.7	2.9	6.7	12.5	20.4
.257 Weatherby	87 PSP	3825	0.7	2.9	6.9	12.7	20.9
.257 Weatherby	100 PSP	3555	0.7	2.8	6.6	12.2	20.0
.257 Weatherby	117 PSP	3300	0.7	3.0	6.9	12.9	21.1
.270 Weatherby	130 PSP	3375	0.7	2.7	6.4	11.9	19.3
.270 Weatherby	150 PSP	3245	0.6	2.4	5.7	10.5	17.0
7mm Weatherby	139 PSP	3300	0.7	2.8	6.7	12.3	20.1
7mm Weatherby	154 PSP	3160	0.7	2.7	6.3	11.7	19.0
7mm Weatherby	175 PSP	3070	0.7	2.8	6.5	12.0	19.5
.300 Weatherby	150 PSP	3545	0.7	2.8	6.6	12.2	19.9
.300 Weatherby	180 PSP	3210	0.6	2.6	6.1	11.2	18.2
.340 Weatherby	200 PSP	3180	0.7	3.0	7.1	13.1	21.5
.340 Weatherby	210 NP	2850	0.7	2.9	6.9	12.8	20.9
.340 Weatherby	250 NP	3180	1.0	4.2	10.1	19.0	31.6
.378 Weatherby	270 PSP	2925	1.1	4.7	11.4	21.8	36.8
.378 Weatherby	300 RN	2700	1.0	4.3	10.2	19.3	32.1
.460 Weatherby	500 RN		1.2	5.1	12.2	23.2	38.8
Rimfire Cartridges							
.22 Short, Standard Velocity	29	1045	1.8	7.2	16.6	30.4	
.22 Short, High Velocity	27	1120	1.6	5.9	13.0	23.0	
.22 Long Rifle, Standard Velocity	40	1150	1.2	4.4	9.4	16.2	
.22 Long Rifle, High Velocity	40	1255	1.5	5.5	11.3	18.9	
.22 Rimfire Viper	36	1410	1.7	6.7	14.1	23.4	
.22 Spitfire and Yellow Jacket	33	1500	1.7	7.2	15.3	25.6	
.22 Rimfire Stinger	32	1687	1.5	6.7	15.4	26.5	
.22 Winchester Mag. (CCI)	40	2025	1.2	5.3	13.0	24.1	
Handgun Loads							
.22 Remington Jet	40 SP	2100	1.1	4.6	11.1	21.1	
.221 Remington	50 PSP	2650	0.6	2.3	5.5	10.2	
9mm Luger	115 FMC	1155	1.3	4.7	10.0	17.1	
.38 Special	110 STHP	945	0.8	3.2	7.0	12.4	
.38 Special	125 STHP	945	0.7	2.8	6.4	11.5	
.38 Special	158 LWM	755	0.8	3.3	7.5	13.6	
.357 Mag.	110 JHP	1295	2.2	7.6	15.6	26.1	
.357 Mag.	125 JHP	1450	1.5	6.2	13.3	22.2	
.357 Mag.	158 JSP	1235	1.5	5.2	10.7	17.9	
.41 Remington Mag.	210 JHP	1300	1.3	5.0	10.3	17.2	
.44 Special	246 Lead	755	0.7	3.0	6.8	12.3	
.44 Remington Mag.	180 SJHP	1610	1.3	5.8	13.2	22.7	
.44 Remington Mag.	240 SJHP	1180	1.1	4.1	8.6	14.6	
.45 Long Colt	250 Lead	860	0.8	3.1	7.1	12.8	
.45 Colt Automatic	185 SWM	770	1.5	6.2	14.3	26.1	
.45 Colt Automatic	185 JHP	940	0.8	3.1	7.0	12.7	
.45 Colt Automatic	230 Metal Case	810	0.7	2.8	6.4	11.5	

*HP, Hollow Point; PSP, Pointed Soft Point; SP, Soft Point; RN, Round Nose; BP, Bronze Point; PP, Power-Point; PP/FP, Power-Point/Flat Point; ST, Silver Tip; FP, Flat Point; JHP, Jacketed Hollow Point; NP, Nosler Partition; FMC, Full Metal Case; STHP, Silver Tip Hollow Point; JSP, Jacketed Soft Point; SJHP, Semijacketed Hollow Point; LWM, Lead Wadcutter Match; SWM, Semi-Wadcutter Match.

The Original Hornet

Joseph B. Roberts Jr.

I N the *American Rifleman* for June, 1930, Army Ordnance Capt. Grosvenor L. Wotkyns published a two-page article describing his experiments with the .22 Winchester centerfire case modified to use .223-inch diameter jacketed bullets. Six months later, Col. Townsend Whelen authored "New Dope on the Wotkyns Cartridge." Finally, in September, 1931, came the announcement, again from Whelen's pen, of "The Standardization of the .22 Hornet Cartridges and Rifles." A classic cartridge had been born—for the second time.

That's correct! The "Hornet" announced in September, 1931, was the *second* cartridge so named and so intended. The first, which predated the Wotkyns-Whelen effort by over 30 years, was announced in a predecessor of the *American Rifleman, Shooting and Fishing,* for March 22, 1894. Before recounting the tale of the Hornet that did fly, it might be interesting to take a brief look at the one that didn't.

The first "Hornet" was devised by Somerville, Massachusetts, gunsmith Reuben Harwood. According to A.C. Gould, then the editor/publisher of *Shooting and Fishing,* who wrote the announcement, Harwood conceived "the idea that there was room for a good serviceable .22 caliber central fire cartridge for small game shooting."

Harwood's solution to the problem as he saw it (since he apparently did not consider the .22 WCF to be "serviceable") was to reduce the neck of the .25/20 single-shot cartridge to accept a .22 caliber bullet. Loaded with charges of 16 to 20 grains of black powder and either 55-grain or 63-grain lead bullets, Harwood's little cartridge and its performance prompted one observer who saw

This article first appeared in American Rifleman

Four cartridges have been named .22 Hornet or inspired it. They are (from left): .25/20 Single Shot, Harwood's Hornet of 1891, .22 WCF, and present-day Wotkyns Hornet.

it to remark, "That's a regular Hornet," and a regular Hornet it was—at least for a while.

By March of '94, when Gould reported its existence, the J. Stevens Arms and Tool Co. had already made known its intention to manufacture rifles for the Harwood cartridge. Similarly, J.H. Barlow of the Ideal Manufacturing Co. had an-

Correspondents and colleagues, campaign-hatted "Grove" Wotkyns, and Townsend Whelen (right) worked together on reintroducing .22 Hornet.

nounced bullet molds and loading tools for it. Harwood, of course, was altering rifles to handle his Hornet just as fast as he could drum up the orders.

What happened to Harwood's Hornet? Why, with such a promising start, didn't it make a better showing? The answer is that it *did*—as the .22 Lovell. But in 1894 it was a different story.

Harwood was limited, for practical purposes, to black-powder loads (though he did experiment with the duplex loadings that were then coming into vogue) and to lead bullets. His Hornet was not *that* much of an improvement over the .22 WCF for small game, and it was never really intended to be a useful target cartridge.

Into the bargain, Reuben Harwood's health failed, apparently seriously, in late 1894. The spells of sickness were not enough to kill Reuben Harwood—he continued to write for *Shooting and*

Fishing (as "Iron Ramrod") and later (as "Aberdeen") for *Arms and the Man* until his death shortly before World War I—but they did kill his Hornet. Without Harwood's sales efforts, the cartridge that might have been great withered and died. Stevens may have built rifles for the Harwood Hornet, but none has yet come to light. The same is true of the Ideal loading tools. Harwood himself may have made up as many as 50 rifles chambered for his cartridge.

And so the scene shifts, from Massachusetts to California, from the 19th century to the late 1920s. A new cast of characters and a new Hornet cartridge step into the limelight.

Capt. Grosvenor L. Wotkyns, Ordnance Department, U.S. Army, was a shooter and experimenter long before he became a soldier. As a soldier he was a bonafide expert on small-arms ammunition. In that capacity he served on the

George Woody used .22 caliber Springfield (top) to build Al Woodworth's first Wotkyns Hornet. Winchester waited to see if Hornet would fly before offering it in Model 54 (bottom).

boards that corrected the long-range firing tables for the .30 caliber M1906 cartridge, then designed the bullet used in .30 caliber M1 ammunition that replaced M1906 Ball.

In 1929 Wotkyns was stationed in California—one has to suppose at Benecia Arsenal. While there, he became interested in filling the same gap in commercial cartridge availability that had inspired Harwood's effort. Wotkyns' approach, however, was a bit different. Wotkyns wanted, and had built, a ''Small Game Express Rifle.''

He used a BSA No. 12 Martini action fitted with an M1922M1 .22 caliber Springfield barrel. George Titherington of Stockton, California, made a reamer—based on the same .22 WCF that Harwood had sought to supplant but with the case neck reduced in diameter so that it would hold .223-inch diameter bullets—and used it to chamber Wotkyns' Martini. He then used the same reamer to make both a sizing die and a straight-line bullet seater.

While Titherington was altering the Martini to centerfire and doing the barrel work, Wotkyns went to work on the bullets. Because available bullets for .22 centerfires generally ran .226-inch diameter in those days, and because Wotkyns' .22 Springfield barrel had .223-inch grooves, Wotkyns' job was to find those bullets which when reduced in diameter to .223-inch would still give acceptable accuracy. This he did, selecting the full-jacket 45-grain bullet made by UMC for its 5.5mm Velo Dog ammunition. He also used a 44.4-grain .2235-inch diameter jacketed soft-point

bullet sold by A.O. Niedner for use in Neidner Rifle Corp.'s cartridges. The Velo Dog bullets were reduced to .223-inch by simply running them through a sizing die, after which Wotkyns loaded them into his supply of necked down .22 WCF cases. Du Pont No. 1204 powder was the propellant of choice, 12 grains with either the Velo Dog bullet or the Niedner.

Trajectory of Wotkyns' ''Express'' cartridge using either bullet was such that a rifle zeroed to print about 1/2-inch high at 50 yards would shoot to point of aim at 100 yards. From the trajectory and the known ballistic coefficients of the bullets involved, Wotkyns calculated that the velocity was between 2,400 and 2,500 fps and that pressures ran about 35,000 psi. Then, satisfied with the success of this efforts, he wrote the article published in the *American Rifleman*. He also wrote to Townsend Whelen.

Whelen picked up where Wotkyns left off. Certain that further improvement was possible, the colonel contacted Al Woodworth in the Proof Services Department at Springfield Armory and Capt. George Woody, also at Springfield. The three—Whelen the idea man, Woodworth the designer, and Woody the man with the tools—set to the task.

Woody started out by making a reamer, the dimensions nearly those of the .22 WCF, but with slight reduction in neck diameter to accommodate the .223-inch bullets that would be used. Woody was not satisfied.

''The chamber neck was slightly large,'' he

Both Al Woodworth (left) and Capt. George Woody tried Hornet (Wotkyns' version) on their local groundhog population before pronouncing it ready for public debut.

wrote later, "and the junction of the cone and neck too sharp." A second reamer corrected these deficiencies and was accompanied by a roughing reamer also used to make a case-sizing die.

Woody then went to work on a rifle, a .22 caliber M1922M1 Springfield owned by Al Woodworth, plugging and redrilling the bolt face and converting the striker from rimfire to centerfire., lapping in the locking lug, cutting a counterbore for the WCF rim, and altering the extractor. Finally he rechambered the barrel, and Woody and Woodworth took the modified rifle groundhog hunting.

The pair used Wotkyns' Velo Dog bullets for those first outings, plus another supply of the same bullet turned backwards and reformed according to Woodworth's ideas to give a 45-grain flat-base, copper-jacket bullet with a round, exposed lead nose. To the chagrin of the woodchuck population around Springfield, Massachusetts, the wildcat cartridge worked every bit as well as Wotkyns had promised.

During the spring and summer of 1930, Woodworth, Woody and Whelen continued efforts at improving the breed. Limited by Du Pont 1204 powder to bullets weighing 40 to 45 grains (2,400 fps velocities using 1204 and bullets heavier than 45 grains required heavily compressed loads unacceptable to the experimenters), the trio set about devising bullet shapes best suited to reten-

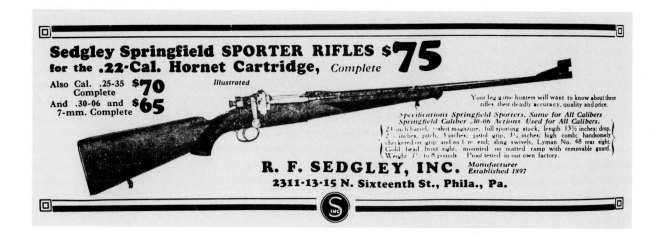

42

tion of velocity. The result was a Woodworth-designed 45-grain bullet with what Whelen—the publicist—described as "a medium-sharp point." Woody made the forming dies and assembled the bullets from lead slugs and trimmed but unheaded .22 Short cases.

Machine-rest groups, fired with these bullets and 11 grains of Du Pont 1204, ran from 2½-inch to 3½-inch extreme spread at 200 meters. Velocity ran 2,400 fps. Drop at 200 yards from a 100-yard zero was reported at about 6 inches and cartridge cases lasted for half a dozen reloadings.

Satisfied with their results, Whelen, Woodworth, and Woody then did two things. They named the cartridge—the Wotkyns Hornet, after their colleague and because the bullet stung like a hornet—and they showed it to officials at Winchester and elsewhere.

Winchester jumped on the new round. By October, 1930, Edwin Pugsley, vice president and general manager at Winchester, had produced and had shown to Whelen factory cartridges for the "Winchester Hornet." A standard cartridge and chamber drawing, published in the *American Rifleman* for September, 1931, is dated the preceding April 13.

Winchester's factory ammunition used a 45-grain bullet after the pattern developed by Woodworth and Woody at 2,350 fps muzzle velocity.

Oddly, Winchester did not introduce a rifle to go along with the new cartridge. Instead it waited, possibly to be certain of the market, until 1933 before adding its own Hornet to the caliber lineup for the Model 54.

Winchester did not have to worry about a market for the cartridge, though. Even before a formal announcement was made, at least two custom gun companies were advertising rifles chambered for the Hornet. In fact, Griffin & Hobbs (earlier *and* later Griffin & Howe) advertised the new ".22 caliber express rifles" in July, 1930, the month after Whelen's story appeared. And, a month before Wotkyns' first Hornet story in December, 1930, Griffin & Howe advertised the .22 Wotkyns Express. R.F. Sedgley's ad in the *American Rifleman* for April, 1931, featured rifles for the .22 Hornet; a cartridge advertised by G&H in January. And, the Stoeger catalog for 1932—its contents derived from products available in 1931—lists Savage Model 23-D rifles chambered for .22 Hornet at a retail price of $32.95.

Over the years since its formal introduction, the .22 Hornet made a lot of fortunes and a lot of friends. Before the .222 Remington sent the Hornet into a temporary eclipse, proponents of the little cartridge had fostered two very successful and popular canister powders and had made careers for at least two men.

The demand by handloaders for a propellant capable of driving a 48- to 50-grain Hornet bullet at Wotkyns' original 45-grain velocities prompted Hercules Powder Co. to develop and produce "2400." Announced in mid-1933, 2400 quickly displaced Du Pont 1204 in Hornet use. Woody loaded 10.6 grains of 2400 and bullets of 36 to 40 grains weight to get 2,650 fps. Hercules, in its advertising, promised 3000 fps using 35-grain bullets and 2400, though it did not list a charge weight. That 2400 worked so well in Hornet (and other) cases that it sounded the death knell for 1204.

Du Pont, however, was not ready to throw in the towel. In 1935, 1204 was replaced by IMR 4227. If there is a powder better suited for the Hornet than Hercules 2400, it is Du Pont IMR 4227. Reloaders today can thank the Hornet for both of these fine and versatile propellants. (Hercules no longer suggests 2400 for the Hornet, but Winchester's 680 more than fills that void.)

The same is true for bullets. Until 1932 or '33, .22 caliber bullets for .22 centerfire cartridges were usually of .226-inch diameter and ran from 60 to 80 grains weight. The large ammunition makers loaded them in the .22 Savage Hi-Power—the "Imp." Custom shops loaded them in proprietary cartridges such as Niedner's .22 Magnum. Unless specially made, as were .22 Niedner Baby Hi-Power bullets that Wotkyns used in his early experiments, there were no jacketed .223-inch bullets to be had. The Hornet changed all that.

By the end of 1933, R.B. Sisk of Iowa Park, Texas, was in the business of making and selling 36- and 40-grain bullets intended especially for the Hornet.

In later years, Sisk's line expanded to include .22 caliber bullets of .223, .224, and .228-inch diameter, and in a full range of weights. But the Hornet bullets that R.B. Sisk went into business to make were always sales leaders.

As it made Sisk's career, so the Hornet made Grove Wotkyns'. When he retired from active military service in 1933, Wotkyns' reputation as

an ammunition expert was already well established. He continued to be active in the field of cartridge design for the rest of his days, devising, among others, the .22/250 (or Wotkyns Original Swift) and the .220 Swift.

Wotkyns, however, earned his retirement livelihood in the same manner as R.B. Sisk—from bullets. Drawing heavily on the Woodworth-Woody experience, Wotkyns designed the series of bullets known as "8-S." Featuring an ogive of eight calibers radius that gave them the appearance, almost, of spire points, Wotkyns' 8-S bullets were first made by J.B. Sweany, then of Winters, California. Later, when arrangements with Sweany fell through, Wotkyns collaborated with Charlie Morse of Herkimer, New York, to sell Wotkyns-Morse 8-S bullets.

If powder companies, the inventor, and the bullet makers prospered on account of the Hornet's popularity, so did the riflemakers. Once Winchester decided to get its feet wet, it waded in right up to the knees. Model 54 rifles chambered for the .22 Hornet began leaving New Haven in 1933. Soon both 20- and 24-inch barrelled standard rifles were available, along with Super Grade, Heavy Barrel, and Target Model rifles. When the Model 70 replaced the 54, Hornets were available in every variation except the target rifle and the bull gun. The last Model 70 Hornets were made in 1957.

For those who, in 1949, did not want to spend $123.25 for a Model 70, Winchester also made the Model 43 in .22 Hornet. Available as a plain-vanilla rifle or in a "Special" version with better wood, checkering, and optional-aperture rear sights, Model 43 prices started at a mere $54.

Savage, the first armsmaker to sell Hornets, continued to do so for longer than any other manufacturer. Savage 23-Ds were made from

1931 until 1947. The Model 219, a top-break, single-shot, joined the line in 1948, lasting until 1965. Finally, from 1948 until just last year, the Model 340 has been made in a variety of useful calibers, .22 Hornet among them.

There were others who built Hornets. Griffin & Howe and R.F. Sedgley have been mentioned as pioneers. There was also H&D Folsom Arms Co., which in the '30s at least, sold Winchester single-shot "Muskets" converted—by Sedgley, some say—from .22 rimfire to .22 Hornet.

Even the United States Armed Forces got into the act, contracting for, purchasing, and issuing at least two "survival rifles" in .22 Hornet. From about 1950 onward the Air Force issued the Harrington & Richardson-built M4, a magazine-fed, bolt-action rifle with a takedown barrel and collapsible wire stock. The M4 was followed by the M6, an over-under .22 Hornet/.410-bore top-break, approximately 66,000 of which were made by Ithaca Gun Co. in 1951 and 1952. A third survival rifle, the 1957-vintage MA-1, was co-designed by the Air Force and Armalite Div. of Fairchild Aircraft Corp. A bolt-action repeater like the M4, the MA-1 could be disassembled and stored in its own hollow buttstock.

Hornets came from overseas, too. Before World War II, Charles Daly sold Hornet rifles made, presumably, in Germany, but which looked suspiciously like Savage Model 40 actions with European-style stocks and set triggers.

Over the years other European gunmakers have joined the Hornet parade. The cartridge was, and is, quite popular with makers of combination guns. Birmingham Small Arms, Ltd., whose Martini action was used to start the whole

thing, made its Majestic, Monarch, and Imperial bolt-action rifles in .22 Hornet. Sako and Walther also introduced Hornets into their respective postwar product lines. Sako's "Vixen" in Hornet went out of production in the early 1970s. Walther's KKJ Hornet lasted until late in the same decade.

Oddly, or perhaps not so, Remington never became involved with the .22 Hornet. Instead, Remington developed the cartridge that almost did the Hornet down—the .222 Remington.

Available as a factory-loaded round, superbly accurate and as effective at 200 yards as the Hornet was at 100, the .222 Remington quickly cornered what had been the Hornet's market. Grosvenor Wotkyns' ".22 Express" nearly died before something happened to change the course of events—something called urban sprawl!

On the East Coast, always the home of the Hornet, cities and towns expanded until their boundaries touched. Whole counties, predominantly dairy farming areas before 1950, became bedroom communities by the late 1960s. The farms, accessible to woodchuck hunters equipped with Swifts and Varminters and a myriad of high-velocity wildcats ideal for long-range shooting, disappeared and were replaced by small plots measured in tens of acres rather than hundreds. The big guns were suddenly too big, too loud, with too great a range. It was time for the Hornet to come home.

It has been over 90 years since Reuben Harwood first devised his Hornet. The cartridge has come and gone, and come back again, and it has done so twice. It was "a regular Hornet," in 1894, and, by golly, it still is.

PART TWO

SHOTGUNS

Develop a Shooting Style

Don Zutz

At the tag end of last summer, I chanced into a round of skeet with one young man who was having all sorts of trouble. He couldn't remember which target came first, the high house or the low house; he had some problems loading his gun; he short-shucked his field-grade pump on doubles; and his foot position and stances were horribly grotesque. His scores were lamentable.

Despite having put on such a bad show, however, he resented it whenever anyone referred to him as a beginner. It seems he had shot a round or two of trap somewhere in the past, and he'd also fired an earlier round of skeet years ago before entering the army. Nowhere in his mind was there room for any thought that he might be a bit wobbly on fundamentals. What he was doing was just fine, he insisted. It would only take a mite of time to put it all back together.

The fact that he had no style was obvious, as was the fact that he had no comprehension of the basics. But his ego got in the way. He wasn't open to advice, and so he blundered along for several more rounds of skeet, scoring poorly,

fumbling with his gun, and getting all screwed out of position when he tried to apply different stances to different stations.

This sort of person happens along all the time, in trap as well as in skeet. In fact, I believe more people get off to poor starts in trapshooting than they do in skeet, because a lot of instructions can be given between stations in skeet whereas the trap line keeps banging away and there is precious little time for instruction (meaningful instruction) between targets. Yet, beginners in trapshooting get into full squads and are swept along with the flow of the game, their mistakes magnifying as they go.

Unfortunately, many trapshooters build their initial "style" of shooting upon these mistakes. Once a person has begun hitting scores in the 20s, he or she is often averse to altering or modifying things such as stance, hand position or grip, body lean, footwork, etc. So, too, are they averse to altering the way they mount their guns, hold their heads, use their eyes, and point over

This article first appeared in Shotgun Sports

the traphouse. In many instances, these initial styles are awkward and ineffective, but because they eventually work to a certain degree, the shooter sticks with them. Fortunately, a few new shooters do quite well despite debatable style which is poorly suited to the proper execution of fundamentals, and that keeps them in the game. But those who simply can't hit a lick with unorthodox stances frequently get frustrated and leave. If they had spent some time trying to learn improved application of the fundamentals in an attempt to develop a sound style, they might well have gotten more satisfaction from the game.

As mentioned above, one of the detriments to improved shooting is egotism. It defeats coaching. A shooter often doesn't like to believe that he isn't naturally great, that doing things his way won't finally bring perfect scores and championships. Indeed, there is a bit of the prima donna in many people. But anyone who wishes to improve as a shooter, as well as beginners who want to start with good form, must think in terms of developing a fundamentally sound style by a study, and an application, of the basics. To accomplish this, the person must respond to coaching; and it isn't all wrong to attend a big shoot and study the style and execution of a few great champions.

Every world-class athlete concentrates on style.

Anybody can climb atop a diving board and jump into the water, but a world-class diver must do it with optimum body control and a precision entry. Any high schooler can swing from the uneven bars, but an Olympic performer must do it with certain control. And anyone can throw the javelin, but to achieve a winning distance requires a well-developed technique tailored for javelin, not baseball or discus. In other words, every sport has its sophisticated techniques, and trap and skeet shooters would do well to accept the fact that good marksmanship on clays also has fundamentals which are needed to produce a sound style. Hunters who tromp to the trap line once a year just before opening day wearing their bird vests, 10-inch boots, camouflage caps, and using their duck guns simply aren't the epitome of trapshooting grace. Nor do they generally win the round against knowledgeable clay-target shooters. There's more to it than planting boot-encased feet, pointing a gun out there somewhere, and shouting, "Okay, now!" as so many beginners do.

When we speak of style, we speak of the manner in which something is done *by an individual*. There is really no such thing as a universal style of shooting. There are only universal fundamentals, and if they are properly carried out, the targets will break regardless of whether the

Is imitation the sincerest form of flattery? Best bet is working toward personal style of learning and then applying rudiments of shotgunning.

Lightning Strikes Twice

AT THE 1950 GRAND AMERICAN, RUDY ETCHEN BROKE 100 STRAIGHT IN DOUBLES WITH HIS NEW REMINGTON 870 AND PETERS AMMUNITION... THE FIRST PERFECT DOUBLES SCORE EVER SHOT WITH A PUMP GUN.

NOW, 32 YEARS LATER, ON AUGUST 8, 1982 AT THE SHREVEPORT GUN CLUB, HE'S DONE IT AGAIN... WITH THE SAME MODEL 870 AND NEW PETERS TARGET LOADS... AS DEPENDABLE IN THEIR PERFORMANCE AS RUDY HIMSELF.

Most coaches today favor upright stance, with swing from waist, no pivoting of ankles and no bending of forward knee. But legendary champion Rudy Etchen achieved his astonishing clay-target scores with lower stance and bent left knee. Personal variations of style are permissible and often desirable, as long as shotgunning fundamentals are properly executed.

shooter is standing with his feet wide apart or close together, holding his leading hand in close or well out on the fore-end, or kinking his leading knee or not. Today, for example, most trapshooting coaches teach a relatively upright stance and instruct their students to swing from the waist rather than pivoting from their feet and ankles. If a beginner bends his leading knee somewhat, the coaches catch the condition immediately and argue against it. The bent leading knee simply isn't in vogue today.

What we must immediately point out, however, is that a fellow named Rudy Etchen uses a bent left knee in his clay-target style, and if that is so wrong, how did Rudy get all those major championships and high averages in both skeet and trap?

Are the modern coaches wrong? Well, no. Is Rudy wrong? Well, no.

What is wrong is thinking that there is only one set way of going about a proper execution of shotgunning fundamentals. Minor variations can indeed be worked into individual styles without causing myriad misses. However, the personal variables cannot interfere with the carrying out of those fundamentals. If a person has such an unusual stance that his start is slow and/or erratic, it is wrong and can't be justified, even if he does break some targets and even if he claims it's okay under the broad banner of individual differences. This thing termed "individual differ-

ences" doesn't justify anything and everything. Individual differences will be tolerated in shotgunning style only if they permit a positive, repeatable performance of the basics.

If a shooter mounts his shotgun so that it doesn't encounter his cheek properly for easy alignment, that can't be accepted as a sound style under the guise of individual differences. The mounting procedure and eye-to-muzzle alignment situation must be revamped to expedite these fundamentals. And, indeed, there are many shooters out there who ignore such needs to revamp their style simply because at one time or another they had a hot day. Don't let a hot streak cause you to believe that everything you're doing is right! Fluke performances occur in shotgunning, too. A far better assessment of one's overall style is consistency. Consistently high scores tend to mean that one is doing the same thing over and over, which is the essence of good shotgunning, and it also means that the important fundamentals are being grooved. So, too, does consistency mean that nothing in a wingshot's style is getting in the way of the necessary fundamental moves.

A lot of shooters try to emulate the style of champions or they stick with the way they did things during their initial rounds of trap or skeet. Both can be wrong, and they often are. If one had to select from the two, it would be better for a beginning or mediocre shot to follow the style

Deep crouch exhibited by trap-shooter in this photo was once fashionable—and may work well for those who learned to shoot that way—but for many it impedes relaxed, fluid swing.

of champions, different though they may be. On his own, a beginner may develop a troublesome stance that blocks smooth, crisp swings or doesn't have a positive follow-through. It may hinder good timing by leaving the shooter wobbly, slow, or too tight. A homebrewed stance can also include a gun position that invites recoil pain.

This writer happens to know of a young shooter who, being a rather lanky fellow, found himself loose when standing upright; he simply couldn't get his long, stringy muscles to control things properly. The salvation, he thought, would be a deep crouch with body lean and deeply flexed knees. His lean was so exaggerated, however, that the gun's butt pad didn't contact fully. The only part of the butt that bore on his shoulder was the top inch or so of the heel, and it was tucked tightly against the fellow's collarbone. The recoil from a steady drumbeat of trap loads not only pounded his collarbone, but it rubbed the skin to the point of bleeding. He looked more like a baseball catcher than a trapshooter in that stance, which he has since modified so that the butt pad bears fully for a more even distribution of recoil energy. It was one of those occasions when a shooter tried to alter his style and it turned out for the worse rather than the better.

There are times when beginners will try to copy a flashy style being used by a hotshot. This could include anything from a rakish lean to a radical foot position or a certain method of mounting the gun. What such copycats don't realize, however, is that the rakish or radical aspect doesn't fit their body's way of doing things. Too, the beginner may concentrate so much on carrying out the rakish or radical element that the important fundamentals are ignored, which is especially wrong! Don't let little flashy aspects of gun handling or stance obsess you. Concentrate instead on the targets and the fundamentals of hitting them.

Exactly how should a shooter develop his or

her style, if not by watching and trying to emulate others? Style is something that shouldn't be consciously engineered. In other words, one shouldn't be overly concerned with how he or she looks. Moreover, one shouldn't think that making the same gun-handling moves and taking the same stance as a champion will inevitably be right. Appearance and copying aren't important factors in developing a personal style. We go right back to the main theme of this piece—namely, that one's personal style derives from perfecting the fundamentals. First, you have to find out how you can do things best; this may take a couple or three seasons of shooting. Then, after learning how your body can apply the fundamentals most effectively and consistently, you have to work to groove these positions and moves, which may take another season of conscientious application. And if somewhere along the line you learn that one of the fundamental factors is being blocked out or affected, a minor adjustment must be made. In trap, for instance, many beginners try the so-called "square stance," which has both feet together with toes pointing toward the trap and shoulders square to the trap. This feels good to many because it gets the butt directly under their master eye and places the pad firmly against their shoulder. But many such shooters find themselves unable to hit the wide rights and/or lefts. Right-handers especially come up short on extreme rights, while lefties come up lacking on wide lefts. There simply isn't any pivot or hip action to handle the wide angles, and some adjustment must be made to free the body for movement to the wide sides. Thus, no matter how you want to look or how you want to stand, there must a consideration for overall technique and the flawless application of fundamentals.

A shooter's own particular style often just plain happens as he or she masters the basics and learns to cope with all target angles. Other parts of the style include comfort and the ability to start smoothly and to follow through.

Equally important is developing a style which

In this photo sequence, Editor Jim Carmichel shows exemplary form as he swings on high, incoming dove. Same fundamentals and shooting style can be applied to breaking clays or taking game.

doesn't tire you out over 100-200 targets. It is a good idea to remain compact and to use a minimum of actions.

The concept of style does imply individual distinctiveness. However, don't go to extremes to be distinctive. Instead, the best bet is working toward a personal style by learning the rudiments of shotgunning and then applying them to perfection. Style for the sake of style alone is nonsense. Indeed, styles are the end result of each individual's method of putting the fundamentals into play, and you'd be far ahead by not worrying about how you look. Worry instead about how you score. Make all adjustments only with a mind toward improved averages, meaning the execution of the shot is more vital than the cosmetics. If a wide stance scores better for you, use it. But if a wide stance loses birds for you, drop it immediately and close the stance. If a high right elbow serves you well, keep it high. If you break more birds with your leading hand out to the fore-end tip, then grip it there and don't worry about Norman, who uses a shorter leading-hand grip.

How do you know when you've arrived at your final style? Your scores will tell you. The proper application of the fundamentals should bring high scores. All this must be evaluated within the parameters of one's basic abilities, of course, but a shooter who averages 89 percent will know he's doing something better if he makes a switch and suddenly runs 94 percent.

In conclusion, then, it must be repeated that style can't be consciously engineered for any given appearance. It must develop naturally according to a shooter's individual abilities to use the fundamentals so necessary to good scoring. The best shooters and coaches differ in what they teach and what they themselves do with a gun, but they all have a full grasp of the fundamentals. In a sense, they're all doing the very same things, only they're dong them with slightly different foot positions, leans, hand placement, and gun-handling moves. But let the bird out, and they all start smoothly, they all have crisp swings that don't check, they all have fine timing on the trigger pull relative to the swing speed, and they all follow through. Thus, style is only the cosmetic appearance of applying the necessary fundamentals, and before a shooter worries about how he looks on the line, he must learn the basics of shotgun shooting and groove them according to individual effectiveness. Let style develop as it will while you perfect the basic elements of sound wingshooting according to your physique and athletic talents.

They're Playing Your Shotgun Game

Jim Carmichel

See if one of these descriptions fits you:
- You are an artist with a shotgun. Your style is so sweet that your hunting pals sometimes stop shooting just to watch the way you bring a gun to cheek and gracefully swing on pintails or pheasants. You are so lightning-quick that a grouse has less than half a chance, and two quail fall out of every covey flush. You like shotguns and you'd love more off-season shooting, but skeet and trap don't offer the challenge you crave. After a few round of either, the numbing repetition begins to pall, and you are distinctly turned off—even repelled—by the colorless, robotlike shooting styles adopted by most of the serious competitors. "If I have to do that to win," you tell yourself, "I'd rather not bother."
- You love hunting anything that flies, but you miss so many shots that most hunting trips are an exercise in frustration. You've tried to improve your form and swing by practicing on the skeet field, but what you really need is a way of preparing for the various speeds, angles, and distances you get with live gamebirds.
- You used to love competitive shooting with a shotgun, but after several years of blasting clay targets, you're looking for something different, something more challenging but at the same time more relaxed. You would like a game that holds your interest and pits you against other top wing-shooters, but at the same time doesn't call for the bang-bang-bang of hundreds of shots per tournament that you've come to dread in organized claybird competitions.
- You're a newcomer to wingshooting games and you want to get involved in clay-target competition. What you've seen so far, however, causes you to hesitate. Skeet looks like fun, but to be a serous competitor you have to shoot four different gauges. That means a heavy investment in guns and related equipment. Also, skeet tournaments are endurance contests, with the spoils often going only to the strong. The people are nice, but the game is so ritualized that it has become a game for clones. You would prefer a game that would allow you to develop your own individual shooting style and in which you would be able to win because of your style.

Trap looks exciting, and the big cash prizes are attractive, but the money appeals to lots of other shooters, too. Some of them get ugly when they don't win. You saw one trapshooter toss his gun into the trunk of his car and drive off after missing his 37th target out of 100. "No chance of winning now," he growled, "so there's no point in wasting shells on the rest of the tournament."

It seems to you that a game that calls for a perfect score just to get into a shoot-off for the money has a couple of problems. It's too easy (or there wouldn't be so many perfect scores), and it becomes another endurance contest. A much more appealing game would be one that offered something for everyone—a game so difficult that no one could shoot a perfect score but with enough variety to let your strong points shine. The game would be so challenging that your main competition would always be yourself, and it would be so satisfying that you'd never want to stop.

If you find yourself in any of these categories, or if you just like to have fun with a shotgun, the game you've been looking for is looking for you. It's called Sporting Clays. In England, just say "Sporting," and everyone will know what you're talking about. In the British Isles, where shotgunning is a way of life for many sportsmen, no other game has had the impact of Sporting Clays or enjoyed such tremendous popularity. Skeet and trap, as we know them, were once

This article first appeared in Outdoor Life

popular at British shooting clubs but now seem to be fading into obscurity. Even Olympic-style shooting games, so popular in Europe, take a back seat to Sporting Clays on the English-speaking islands.

Whenever I discuss Sporting Clays with American shooters, they often ask if it is similar to any of our tricky shotgun games, such as Crazy Quail. Any similarity is only accidental because Crazy Quail and the like are designed to be as difficult as possible. A Sporting Clays circuit, by contrast, is designed to duplicate the flight of gamebirds. American skeet, it should be pointed out, was originated to approximate the flight angles a hunter might encounter when hunting quail or grouse. But the angles on a skeet field represent only one kind of hunting; a Sporting Clays layout represents several kinds of game. On a single Sporting Clays course, you are likely to shoot at a target skipping along the ground like a rabbit, toss shot at a pheasant 40 yards overhead, and try to catch up with springing teal that rocket skyward at nearly vertical angles. Other colorfully named stations give you an idea

Drawings by Ken Laager

The Tower: *Stripped of scorekeeper and wire barrier that prevents firing in unsafe directions, low-tower situation is shown in drawing. Many fields have higher tower that is useful in simulating flight of passing waterfowl. Photo shows Clare Conley, Editor of* OUTDOOR LIFE, *getting set for doubles from tower on English field. Clays mimic ducks or high upland birds. When shooter calls "Ready!" traps are sprung at once or up to three seconds later.*

of what to expect: "Woodcock pair, rocketing pheasant, quartering moorhen, fur and fowl, grouse butts, dropping duck, crossing partridge."

Wingshooting games, for either practice or competition, have been around almost as long as there have been shotguns. More than 200 years ago, blue-blooded sportsmen were blazing away at live pigeons released from traps. (Yes, that's how modern clay-target launching machines got their name.) The betting was heavy. By the 1880s, glass balls launched from mechanical traps had largely replaced live pigeons, and the top marksmen were national heroes. Glass targets caused some obvious problems, so some enterprising fellows came up with a target made of baked clay. With these inexpensive targets, there was no limit to the games that could be played with a shotgun. In America, trap and then skeet became the two favorites, each governed by a ruling association that establishes the rules and oversees formal competitions.

In England, a single association, the CPSA (Clay Pigeon Shooting Association) was formed to organize and promote all forms of clay-target competition. That was in 1928, when trap was more popular than either Sporting Clays or skeet among British shooters.

During the early 1970s, Sporting Clays took off in a big way and became the runaway favorite. Why it has become so popular is not at all hard to understand once you give the game a try. It is even fun to watch! Today, Sporting Clays dominates the British shotgunning scene so much that skeet and trap are simply referred to as the "other" games. Otherwise, if you're on English

Fur And Fowl: *Gunner swings on "bounding hare," actually a claybird rolling on edge. Ground is rough, so "hare" bounds with utter reality. When shooter fires, "angling grouse" is launched "on report," as the English say. This is common situation in English hunting. Drawing shows setup, but in real life, "trappers" (pull boys) would be protected by barriers that are usually made of bales of straw.*

Woodcock Pair: *Two clay birds, launched together, imitate low-flying woodcock. Some gunners deliberately try to "kill" both birds by firing one shot between them. If you make this difficult hit, it counts.*

soil (or in Scotland or Wales) and you're talking about competitive shotgunning, you're talking about "Sporting."

From the end of World War II until the mid-1960s, membership in the CPSA trudged along with a growth rate that was comparable to those of the respective trap and skeet associations in the United States. But then Sporting Clays took off like a rocket. To give you an idea of how big it has become, consider that there are seven times as many CPSA members today as there were in 1970 and that there are some 25 times more shooting clubs now than there were a generation ago.

These numbers are amazing because they are such a dramatic contrast to the dismal growth rates of organized clay-target shooting in the United States. Consider further that during the first half of the present decade, sales of shotguns—indeed, all guns—declined to the lowest point in memory. During these same years, British gun salesmen, profiting from the Sporting Clays boom, were sporting around in expensive motorcars and, in some cases, having to ration their more-popular guns.

A further contrast dates back to the 1970s, when the oil shortage and high gasoline prices created a serious drop in American clay-target shooting. At the same time in England, participation in Sporting Clays was showing the fastest growth rate ever. One other interesting contrast is the dependence of American-style trap and skeet on reloaded ammunition. The blooming of both games here occurred during the 1960s, when dependable, inexpensive, and easy-to-operate reloading tools became available. Since that time, participation in U.S. wingshooting games has been closely linked to the shotshell-reloading scene. When the price of shot spirals upward, as it did a few years back, participation in U.S. trap

or skeet goes down. In England, there are very few handloaders, and virtually all Sporting Clays are shot with new factory loads. What these various comparisons tell us is that the popularity of Sporting Clays is not riding on a crest created by economic circumstances, and its popularity cannot be attributed to any other source of outside stimulation.

In fact, rather than depending on a bountiful economy for its growth, Sporting Clays is contributing to the British economy. In addition to land-office gun sales and the related growth in ammunition sales, accessories, and gunsmithing services, there is also a boom in commercial shooting grounds, where coaches and their assistants make a very comfortable living by teaching the finer points of wingshooting and by refining the skills of Sporting Clays competitors.

In the United States, where successful commercial shooting grounds have been about as rare as land-walking whales, two Sporting Clays layouts are not only paying for themselves, but also furnishing a very nice livelihood for their operators. The Highland Bend Shooting School, operated by Jay Herbert, and the Champion Lakes Gun Club, run by Danny McMillan, are both located in the Houston, Texas, area, and both are successful because they offer Sporting Clays shooting and solid coaching. Jay Herbert summed it up when he said that by using a Sporting Clays layout for coaching, he helps his students to become very proficient game shots. Another commercial operation (Beretta U.S.A. P.G. Shooting Center, 10400 Good Luck Rd., Glenn Dale, MD 20769) opened in the East, and several clubs, such as the Wilderness West Club in Mississippi and the Minnesota Horse & Hunt Club, are offering Sporting Clays shooting. I expect there will be a rash of new Sporting Clays operations during the next few years.

How is it that a backwoods game has become the overwhelming favorite in a nation of sophisticated wingshooters and stands on the brink of enormous popularity in the United States? The simple reason is that the game is fun. Fun! The people are fun to be with, and you go home feeling like you had fun.

The fun starts with the Sporting Clays layouts. Trap and skeet courses strive to be identical; Sporting Clays fields are remarkable for their individuality. I've tried the game on three different layouts in the U.S. and a half-dozen in England and Scotland, and no two have shared more than a superficial similarity. In this respect, they resemble golf courses because they are designed and laid out to take maximum advantage of the natural contour and beauty of the land. Rather than leveling hills and cutting trees, as is often necessary for other forms of shooting, Sporting Clays layouts are simply incorporated into the existing terrain and cover.

The only critics of Sporting Clays I've ever encountered have never given the game a try. They complain about the differences in layouts and feel the game can never catch on until all fields are alike, as they are in trap and skeet. That is as unreasonable as claiming that golf cannot be successful until all courses are identical. One of the reasons Sporting Clays is never boring is that it is never the same. The layouts differ, and most layouts are altered from time to time so that they offer fresh challenges. Another reason Sporting Clays competitions are so popular is that most shooters are always eager to try a different layout and different shooting situations.

It is almost as difficult to describe the way the targets fly on a Sporting Clays course as it is to describe the course itself. The traps are situated and angled to make maximum use of existing trees and land contours. The rule that prevails when laying out a Sporting Clays course is that the targets must be thrown to duplicate shots you would encounter when hunting.

If the goal of Sporting Clays were simply to make the shooting as difficult as possible, it would be easy enough to either throw the targets so high or so wide that they would be out of reach, or launch them too fast to be seen. Instead, a well-designed course beats you with deceptive angles and fools you with illusions. When you miss a target a target at Sporting Clays, you miss it for the same reasons you miss a duck or a dove or a pheasant—because you misread its speed and angle. The hardest shots are often the ones that look the easiest. They don't try to fool you with surprises, either. If you don't know which way the bird will fly, you can ask for a look at a flying bird before you "shoot for record."

The one feature that all Sporting Clays layouts have in common is a tower-mounted trap. Loved or hated by shooters, it duplicates the high overhead flight of ducks, geese, or pheasants. Some layouts may have two or even three towers of different heights. After trying to hit a string of targets that are 120 feet overhead, it's not at all difficult to understand why you miss high doves and ducks. And once you get the hang of hitting targets from a high tower, you're going to be a lot better in the game field.

Overhead: *Common on English layouts is overhead station, where gunner faces away from trap. Clay birds simply appear in air, as is common with wood pigeons or ducks from behind blind or shooting stand.*

Some clubs go to delightful extremes in duplicating field-shooting situations. The British clubs I've visited have authentic-looking stone and sod "grouse butts" that lend atmosphere to the event and give the shooter the feeling of being on a Scottish moor. American layouts such as Champion Lakes and Highland Bend have a "duck blind" in which you shoot while sitting. The realism is wonderful.

Though stone grouse butts and steel towers are usually permanent fixtures at a commercial Sporting Clays layout, some of the other shooting may be impromptu. These temporary shooting stations are another intriguing aspect of the game because they often represent a fresh idea or a new twist to an old angle that the shooters haven't encountered before. A shooter who has been practicing for days and thinks he has the course pretty well figured out may discover on the day of the big tournament that some positions have been changed. This puts out-of-towners on an equal footing with the locals. Most traps on a Sporting Clays are inexpensive hand-cocked affairs mounted on sleds that are easily shifted about. The "trappers" (the trap loaders) are usually shielded by bales of straw.

Bales of straw are also often used to build temporary blinds for the shooters. Whenever possible, the trap is situated so that its exact position can't be seen by the shooter. Ideally, the trap is hidden behind a hedgerow, in some brush, or beyond a tree. This makes the target's flight look more like that of a gamebird. A particularly beautiful arrangement is having the targets launched from behind a row of trees so that when you first see them, they are at the end of their trajectory and gently falling like ducks dropping onto a pond.

The flexibility of Sporting Clays allows an informal group of shooters to organize a once-only competition simply by setting up a few portable traps on a suitable site. At the end of the shoot, everything is carted away. A hilltop can serve as a substitute for a tower. Some clubs even have collapsible towers mounted on trailers. Here in the U.S., there are endless opportunities to set up similar temporary Sporting Clays layouts. The only limit is one's imagination.

The flexibility of Sporting Clays also allows any number of stations. A major club or competition may have 10 stations, but a smaller club may have only three or four. This allows a club to start out small with a minimum investment and expand in its own good time. A very challenging Sporting Clays field can be established for only a fraction of the cost of a single trap or skeet field, and it doesn't require nearly as much work or maintenance.

By now you're probably wondering what a

Springing Teal: *Teal can fly almost straight up, as every gunner knows, and that's what "Springing Teal" do in Sporting Clays. Simplified drawing shows situation without the usual wire safety barrier.*

Sporting Clays shoot looks like and how it is run. The best way to explain is by describing the 1986 British Open Championship. The event was held in mid-May, when a gaggle of American shotgunners from the firearms and outdoor-publishing businesses were invited to go over and see what it was all about. Clare Conley, Editor of OUTDOOR LIFE, and several other editors were members of the group. The tour was arranged by Jim Moore, president of the U.S. Sporting Clays Association, with a lot of help from our counterparts in England. Dan Arms, the Danish makers of high-tech shotshells, and Gunmark Ltd., a leading firearms importer in the United Kingdom, supplied jovial hospitality as well as shells and shotguns. All we had to do was prove we could use them.

Our schedule called for a day of coaching by Mike Reynolds, a champion shot and one of Britain's top coaches, at his Mid-Norfolk Shooting Grounds. The next day, we got another tune-up at the posh Holland & Holland Shooting School.

Then, ready or not, we were to jump in headfirst and take on England's best on the final day of the British Open at the West London Shooting Grounds. The day turned out be typically British, with a cold drizzle leaking out of a blanket of fog.

The British Open Championships course called for 10 shots at each of 10 stations. Most of these were five pairs of "doubles on report," meaning that the second bird of the pair was launched just as the shot at the first bird was fired. At other stations, such as the woodcock station, a low-flying pair of targets was launched simultaneously.

Other combinations that may be encountered on a Sporting Clays course are split pairs, where the targets are separated and fly in widely different directions, or "flighted" pairs, where the targets fly in a close formation like ducks or doves. These flighted pairs are fun to shoot, but can really mess up your mind, especially when targets of different sizes are used. In Europe, a number of claybird sizes are used, ranging down to the "mini," which is about ½ the diameter of U.S. clay targets.

One of the more relaxed and sensible features of a Sporting Clays competition of this type is that you don't have to fire in any particular order. If you don't feel like shooting the first station first, you simply begin at any station you choose. This also has the good effect of dispersing the shooters over the entire course, and it keeps waiting lines to a minimum. This informal arrangement also gives you a chance to figure out the target angles before taking your turn.

The first station I tried was certainly one I should have avoided. It was called "teal and crossing moorhen," and it offered you a fast-rising teal followed, on report, by a wide, near-right-angle target launched at about half again the velocity of a normal skeet target.

Next was a "walk-up," in which the shooter strolled down a hedgerow-bordered path. Crossing targets came out of one hedgerow and disappeared into the opposite one. At another station, a hedgerow-lined lane offered a pair of woodcock thrown together, one flying just above the other. The better shooters could break them both with one shot.

The next station, called "fur and fowl," was utterly unlike any other clay-target setup in that one of the targets, called the "bounding hare," was launched so that it skipped along the ground like a rabbit with its afterburners lit. This required a special type of extra-tough clay target, of course. The shooter's problem was whether to shoot it on the bound or when it hit the ground. In any case, there was no time to congratulate yourself for a hit because "on report" you were presented with an angling-away "flushed grouse." This was a fun-station for spectators because the "bounding hare" made some crazy moves.

While all this was going on, there was lots of shooting at the other stations. One of the tricks to designing a Sporting Clays layout is to position the various stations so they don't interfere with the others. On several stations, the shooter stands inside a wire cage that limits the angles at which he can fire. This is a smart safety feature, and it permits locating the stations closer together than would be safely possible if unrestricted gun swinging were allowed. Because Sporting Clays is such a popular spectator sport, much consideration must be given to crowd safety, especially in that the crowd tends to flow along with the shooters as in a major golf tournament. Whoever heard of such a thing with other forms of target shooting?

The "grouse butts" offered an incomer—low and fast—followed by a going-away target that called for some fancy footwork.

The high tower completely baffled the American shooters. I don't remember anyone hitting more than three or four. All shots were at incoming angles, and they looked easy. This kind of shooting takes a lot of practice and superb form.

Form, by the way, is the name of the game as seen by British shooters. When one misses a target, it is because of the poor form, they believe. American shotgunners tend to deal with a flying target in terms of measurable lead. For example, when we were trying to hit the high-tower targets, our shooting was based on trying to figure out how much we needed to lead or shoot ahead of the birds. A well-coached British shooter, by contrast, would handle the same target in terms of the speed of his swing. It's all a matter of form, they tell you, and once your form is right, you won't miss many targets, live or clay.

A lower tower launched a pair of targets from behind a row of high trees. When the targets came into view, they appeared to be standing still, but were actually zooming down at a fast clip. After a couple of variations of these tower-launched birds, it was nice to finish with a series of uncomplicated crossing targets not unlike those encountered at the middle skeet stations, only higher and wider.

The American rules provide that the gun is to be held off the shoulder, with the top of the stock 4 or more inches below the armpit, when the shooter is ready to call for a target. After the shooter calls for the bird by saying, "Ready!" the trap throws the target immediately or at any time up to three seconds later. Cartridges cannot contain more than 1⅛ ounces of shot, and only shot sizes No. 6 to No. 9 (American designations) are

Woodcock: *Gunner on wood-cock stand gets set for bird. Shot-proof barrier shields trapper and also prevents shooter from seeing when claybirds leave trap.*

legal. In both England and the United States, many shooters do use two different sizes; for instance, No. 6 or No. 7½ for long-range tower birds and No. 9 for incoming, low-flying birds. Many English rules are not specific. For instance, in the British Isles, the shooter is allowed to mount the gun before calling for the target, but most shooters do not.

When the competition ended, all of us would have fired the whole course again if that had been possible. We thought we had done poorly, but by comparing our scores with the others on the scoreboard, we discovered that we had done somewhat better than the norm. In case you're wondering, the American squad, whose members included some well-known skeet and trap champions, posted scores that mostly ranged from the low to high 60s out of a possible 100. The new British champion posted a 90! There was no need for a shoot-off.

Though there are several top shooters who always have the potential to win a tournament, it is hard to pick a winner, and repeat winners are very rare. This is because it is so hard to put it all together on a given day. That's another beauty of Sporting Clays—an unknown shooter, likely or not, can get on a hot streak and be the big winner. Every shooter can win, but win or not, you'll have a hell of a good time and you'll learn a lot about using a shotgun.

Nearly every gun you see at a Sporting Clays tournament is an over/under. Beretta is a favorite along with Winchester and Browning. One sees an autoloader now and then, but there are surprisingly few side-by-side doubles. In fact, there was a separate prize category for competitors who used side-by-sides at the British Open. Those who do shoot side-by-sides, as you would expect, shoot them very well. I've seen more side-by-side doubles fired at American Sporting Clays shoots than I have in England on Sporting Clays layouts.

The best shotgun for you to use when you first try this exciting game is your favorite hunting gun. Barrel lengths in the 27- or 28-inch range are most favored, as are chokes ranging from Improved Cylinder to Improved Modified. Ordinary Modified, or what the British call "Half Choke," is a good choice.

Whether Sporting Clays becomes as popular here as it has in England remains to be seen. There is little doubt, however, that it will be a major force in U.S. shooting and gunmaking. The U.S. Sporting Clays Association is well organized, and it is run by dedicated shooters such as Jim Moore and Bob Davis, both of whom know how to take an idea and run with it. Both have already put more of their time and resources into the game than I have ever known any other individuals to invest in a shooting sport.

Sporting Clays is here! For more information on membership and for tips on how to establish a Sporting Clays club — or even how to adapt existing shotgun clubs for Sporting use — write the U.S. Sporting Clays Assn., 111 N. Post Oak Ln., Suite 130, Houston, TX 77024.

Browning Auto-5

Col. W. R. Betz

Business was good on the morning of May 4, 1898, at the Browning Brothers sporting goods store in Ogden, Utah. In addition to a brisk walk-in trade, the company enjoyed a lucrative wholesale business in neighboring towns in Utah, Wyoming, Nevada, and Idaho.

George Browning, one of the five "Browning Brothers," who functioned as company secretary when he could spare the time away from the shop and sales counter, had just dispatched a letter advising the Neponsit Land and Livestock Company of Evanston, Wyoming, that its order for "2M of the 5½-inch cannon cracker" had been shipped, and acknowledged recipt of an order for a thousand cartridges for Winchester rifles, caliber .44.

The shelves of the largest store of its kind in the area were well stocked with merchandise as illustrated on the company's fancy letterhead — boxing gloves and punching bags; tennis, baseball, and golf equipment; cameras, fishing tackle, bicycles, and guns. Bicycles were the bread-and-butter items, with guns ranking a close second.

One of the guns featured prominently on the letterhead, a long-barrelled automatic pistol, was actually the only one in existence in 1898. It was invented by John M. Browning, the senior member of the firm, three years before and sold to Colt, but it had not been put into factory production.

With Colt's agreement, John M. had sold to the Belgian firm of Fabrique Nationale d'Armes de Guerre (or National Military Arms Works) the rights to make and sell this pistol and a smaller version in all markets outside the U.S. Production of both models began in 1899; the larger Model 1899 failed to gain acceptance in military trials in Belgium and France, but the smaller "pocket pistol," known as the Model 1900, was an instant success all over Europe.

While George Browning was taking care of the firecracker business, things weren't going so well at John M.'s workbench back in the shop. The brothers could always tell when something was bothering John.

When all was going well, he would chat with his brothers and customers and would often hum or whistle while he worked on some new gun design. But when an idea wouldn't jell or the sample pieces of a new gun didn't fit together right, John M. became moody and absent-minded, as he had been for many mornings that year.

In 1889, John M. had discovered that the recoil power generated by cartridge ignition could be harnessed and used to operate the actions of semi-automatic pistols and full-automatic machine guns. Between 1890 and 1897, he had invented and sold to Colt six successful semi-automatic pistols and the gas-operated Model 1895 Automatic Machine Gun, the "potato digger" used by the U.S. Marines during the Spanish-American War and in the Chinese Boxer Rebellion.

John M. next turned his attention to the creation of a semi-automatic sporting arm based on the recoil principle. He had sold two shotgun designs to Winchester, the lever-action Model 1887 and the slide-action Model 1893, and no doubt had Winchester in mind as a customer for the new "automatic" shotgun. John M. had proven the practicality of the basic concept with his pistols and the machine gun, but designing a smaller, lighter gun that would fire an entirely

This article first appeared in American Rifleman

1898 view of Browning Brothers sporting-goods store in Ogden, Utah, shows that the Brownings catered to bicycle as well as firearms trade. John M.'s workbench was located in the rear of the shop. (Ogden Union Station photograph.)

different kind of ammunition presented difficulties that had slowed work on the new gun.

John M. used a toggle action on his first model. The action was opened by pulling upward on a protrusion behind the center of the hinged operating lever, retracting the bolt and cocking the concealed hammer. Somewhat resembling the Luger pistol action, this design had several inherent faults and was soon rejected.

The second model had a closed breech, protecting the action from weather and debris. The bolt had no operating handle; instead there was a lever hinged at the bottom of the buttstock near the toe and attached to a rod inside the stock connected to the bolt. The bolt was opened by pulling back on the stock lever. This model, too, survived only as a museum prototype specimen.

The third and final model had most of the features of today's Browning Auto-5, but it had taken its inventor almost a year to solve one of the most difficult problems he had ever encountered with any of his inventions. The problem stemmed from the fact that smokeless powder had recently been introduced, replacing black powder in rifle, pistol, and shotgun ammunition.

The new powders generated much greater chamber pressure and recoil than had the old black powder. The difference was so great that some older shotguns, particularly those with the lovely but fragile Damascus barrels, had been blown up by the new loads. There was still a demand for black-powder shotgun shells, and John M.'s new automatic shotgun had to be made to work with both kinds of ammunition.

The muscle of the new gun's mechanism was

in the recoil spring, which stored energy from the fired shell and returned the gun to battery after each shot. A spring strong enough to work well with smokeless-powder shells would not allow the action to function properly when a black-powder shell was fired. Conversely, a spring suited for use with black-powder was so weak that parts of the action were subjected to battering and potential failure from the unrestrained force of the recoil from smokeless shells.

John M. and his brother Matthew S. were anchormen on the "Four Bs" trap team (with Archie P. Bigelow and Gus L. Becker) which had won national recognition in the 1890s. They had fired thousands of rounds of 12-gauge ammunition in their Winchester Model '93 shotguns, and John M. was not ready to declare his automatic shotgun ready for reproduction until he was sure it could duplicate that reliability.

The new gun was introduced to John M.'s shooting friends at the Ogden Gun Club, where it was fired hundred of times and functioned satisfactorily as long as only one type of ammunition, black powder or smokeless powder, was used.

After months of test firings and frustration, he finally hit on the solution to the recoil-control problem. It consisted of providing for adjustment of the strength of the recoil spring by adding to the spring assembly a set of friction rings, split and tapered, which could be switched to increase or decrease friction on the magazine tube to suit the kind of ammunition to be used. This simple but effective device outlived the use of black powder and serves the same purpose today, making possible the use of light and heavy loads in the same gun.

In a letter dated March 6, 1899, John M. told T. G. Bennett, vice president and general manager of the Winchester Repeating Arms Company, that he had an automatic shotgun ready to demonstrate. Bennett had bought the first Browning gun, the Single Shot Model 1878, and all other long guns invented by John M. since 1883. Browning's designs were so superior to all others on the market at that time Winchester enjoyed a near monopoly in the sporting rifle and shotgun markets.

But later that month when John M. brought his prototype to New Haven, neither Bennett nor the Winchester technical staff were particularly happy to see it. Winchester management was sold on Browning's lever and slide-action guns which had been so successful over the past 17 years.

The technicians and production managers were not convinced of the practicality of the new idea or its reliability. Neither side of the Winchester house liked the prospect of tooling up a new production capability to make a gun that might turn out to be a flop, nor did they want to risk losing sales of the Browning-designed Model 1897 slide action, the most popular repeating shotgun in the country at that time.

John M. left the new shotgun in New Haven and returned to the Ogden shop to work on more pistol designs, putting the automatic shotgun out of his mind. In the past, he had taken 44 long guns to Winchester, whose expert legal staff had taken on the chore of patenting them in Browning's name. The company had paid cash for all of the guns without hesitation, even those which

were obviously impractical, just to ensure that no competitor got a new Browning-designed gun.

But after several visits and the exchange of sometimes acrimonious letters, Bennett refused to accept John M.'s offer to sell the new gun to Winchester on a royalty basis, and they parted for the last time to become competitors instead of business partners.

On January 8, 1902, John M. called Marcellus Hartley, president of the Remington Arms Company, and made an appointment to demonstrate the automatic shotgun that afternoon. After an hour's wait in the Remington office, John M. was

DEVELOPMENT OF BROWNING AUTO-5

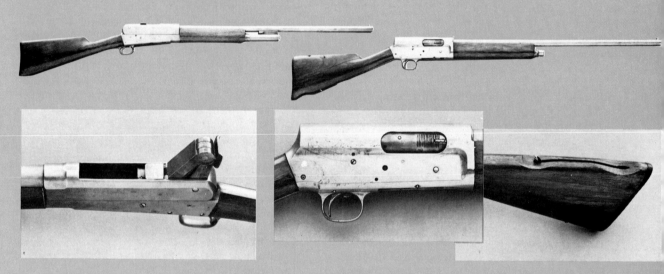

Browning's initial prototype for his automatic shotgun employed toggle action (above left) but inherent design faults caused him to reject it quickly. Second model had closed breech. Its bolt was opened by pulling back lever located at bottom of buttstock (above right). Third model (below) resembles modern Auto-5. Shooters switch from heavy to light loads by adjusting friction ring system designed to handle both smokeless and black-powder loads.

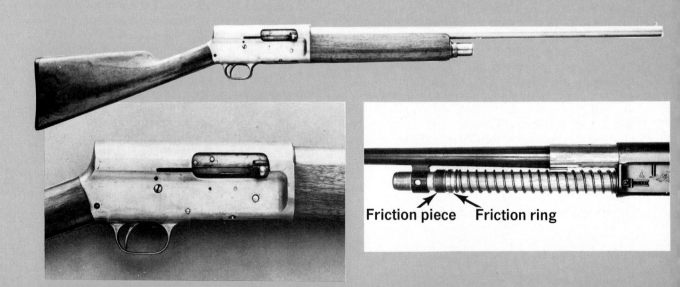

Friction piece   Friction ring

told that Hartley had just died of a heart attack. This second disappointment may have moved Browning to abandon attempts to sell the gun in the U.S. In any case, later that month he sailed to Belgium, the prototype shotgun packed in the huge, round-topped trunk which would see some 60 Atlantic crossings before John M.'s death in 1926.

Henri Frenay, the Director General of the Fabrique Nationale, gave John M. a royal welcome upon his arrival at the plant in Herstal, a suburb of Liege. Sales of John M.'s Model 1900 pocket pistol had exceeded all expectations; the pounding of drop hammers and the slap of a forest of drive belts on the production line saluted the beginning of an era of prosperity which would eventually make the F.N. the largest small-arms maker in the world.

The automatic shotgun had been given a cool reception at Winchester, but it was greeted with enthusiasm at Herstal. Within 60 days, John M. had signed a contract giving F.N. exclusive manufacturing and sales rights to the new gun worldwide outside the U.S. He stayed at the Herstal plant for three months, assisting in the task of transforming a protype into a production piece.

During the process, John M. made several minor changes in details of the design, mainly to facilitate mass production, and the final version went into production in the summer of 1903. The first Automatic 5-Shot 12-gauge gun, Serial No. 1, was shipped on September 17, 1903, to Shoverling, Daly and Gales in New York, the jobber that had supplied sporting goods to the Browning Brothers stores.

Before leaving Herstal, John M. had placed an order for the first 10,000 guns to be shipped to the New York firm, partly, at least, to demonstrate to F.N. his faith in the new gun. The entire lot was sold within a year, somewhat to Winchester's consternation.

F.N. also delivered Auto-5 guns to its European agents in 1903, but the reception of the new gun among Continental shooters was lukewarm. For one thing, the 12-gauge had never been popular in Europe where the 16-gauge was the traditional field gun.

Recognizing this situation, F.N. scaled down the Auto-5 to 16-gauge, and by August, 1909, shipments had commenced to both U.S. and European customers. However, another intangible but nevertheless powerful factor had great effect on sales of the automatic shotgun in both gauges in Europe.

In America, even as late as 1900, hunting was almost as much an occupation as a sport. The American hunter was a pragmatist whose primary objective in the field was to put meat on the table. It was not difficult to convince him of the advantages of a fowling piece capable of firing five fast shots without reloading and that could be depended upon to function without failure under the most adverse conditions.

But in Europe, the chase was still a royal sport indulged in only by the privileged few and conducted decorously with rigid rules of conduct, costume, and weaponry. The advent of the breech-loading double shotgun had been accepted by the gentry, but the American automatic raised eyebrows among shooters and beaters alike. The new gun was regarded as a shooting machine, not a proper fowling piece for a gentleman.

The F.N. salesmen, however, found a ready market for the Browning automatic in many other parts of the world. It became popular as a multipurpose arm capable of downing sizeable animals with buckshot or slugs. Testimonials in the

Each new Auto-5 at Fabrique Nationale was fired five times for function testing (below left) and two more times to check pattern. After getting cool reception at Winchester, Browning's new shotgun was enthusiastically received at Fabrique Nationale, and the Belgians quickly signed exclusive contract to manufacture and sell Auto-5 worldwide outside of U.S.

F.N. files from new Browning fans tell of leopards, gazelles, and wildfowl brought down by the Auto-5 in Africa; deer in Venezuela; geese in Austria; and hares and ducks in Italy. By the 1920s, the Auto-5 had also found favor on the target ranges. The 1921 Grand Prix de Monte Carlo live pigeon shoot was won by M. Lafite with a Browning.

The world record in the 1924 Paris Olympics was set by M. d'Heur of Herstal, shooting an Auto-5. The Italian champion, Guiseppe Cavaliere, used a Browning in his 1926 win over clay pigeons. On the other side of the world, the winning three-man team used Browning Auto-5s at the claybird shoot held by the Kobe Shooting Club, as did the Japanese claybird champion, M. Takahshi, in 1925.

All of John M.'s many firearms inventions have been noted for their simplicity, reliability, strength, and ease of manufacture. In addition to these basic qualities, Browning arms have always been known for the quality of their workmanship. This has not been an accidental virtue originating with the manufacturer, but is the result of John M.'s perfectionism and a tight quality-control policy still enforced by the Browning Arms Company.

At the outset, John M. personally supervised the establishment of inspection and testing procedures which would ensure the uniformly high quality of every Browning Auto-5 produced. Eventually, more than 2,000 separate inspections were performed at various stages between some 650 machine operations. More than 1,500 precision gauges were used to check parts dimensions at various stages.

As required by Belgian law, each Auto-5 was proofed with a "blue pill" overpressure load at the Liege proof house. Each gun was also fired five times at the F.N. plant for function testing, and twice more to check the shot pattern. Barrels that did not meet the firing or the patterning tests were rejected. For many years, pattern targets were supplied with each Auto-5 gun sold, but the practice was discontinued at the request of Browning dealers whose customers insisted on unpacking all the guns in stock in search of the one with the best pattern!

When John M. Browning first visited Herstal in 1902, he found an arms factory that had been established by a consortium of Liege arms makers in 1889 to make Mauser rifles for the Belgian army. The factory had fallen on hard times when the rifle orders were completed and had turned to making motorcycles, bicycles, cars, and ammunition.

The Browning M1900 pistol had started wheels turning again in the firearms plant, and now the Auto-5 would create the need for even more buildings, an expansion that would continue for years as more and more Browning pistols, rifles, and shotguns entered the F.N. Browning line of quality firearms.

Some of the first 10,000 12-gauge Auto-5 guns were sold by the Browning Brothers stores in Ogden and Salt Lake City, Utah. No records can be found to show how many and when they were all sold, but a faded copy of the 1914 Browning Brothers Catalog No. 50 lists the "Remington Autoloading Shotgun, Browning's Patent" at $30, but does not list the F.N. model.

In 1905, with F.N.'s concurrence, John M. had granted the rights to make and sell his shotgun in this country to Remington for marketing under its own name as the Model 11. The agreement gave Remington exclusive rights to sales in this country at that time, but by 1923 John M.'s patents on the Auto-5 had expired, and Browning was again importing the Belgian Auto-5. The older F.N. Auto-5s can still be found in the West, some of them marked "BROWNING'S ARMS COMPANY, OGDEN, UTAH."

Others of these oldest of the Auto-5s may be found with a variety of barrel markings. The first known is "BROWNING AUTOMATIC ARMS COMPANY, Ogden, Utah, U.S.A.," the name of a nonexistent company which also appears on the 1903 catalog which first advertised the gun. No explanation for the name has ever been found. It is quite likely that John M. pulled it out of the air when he placed his first order with the F.N., and it is possible that he had forgotten all about it before the guns arrived.

The 1903 catalog described the "Browning Automatic Solid Breech Hammerless Repeating Shotgun" in several variations. The "Regular

The 2 millionth Auto-5 came off line in 1970, nearly 70 years after first gun was sent from Herstal.

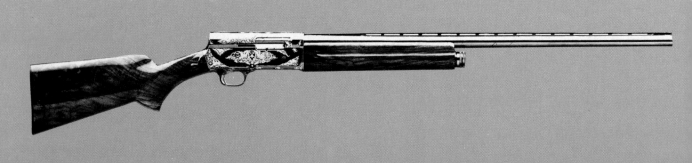

Gun" was offered with straight stock, matted receiver, and 28-inch barrel. The "Trap Gun" was identical except for checkering on buttstock and forearm. The "Messenger Gun" came with a 20-inch barrel, Cylinder bored. The "Two Shot Gun" was just that, a "Regular" model with a smaller magazine.

A deluxe version with select English walnut stock, pistol grip, and checkering was available on special order for an extra charge of $9. Full-choke guns were normally shipped unless the customer specified Modified or Cylinder, the only other two chokes made. Browning recommended the use of 1¼ ounces of No. 8 shot, which the catalog said, "...will put better than 340 pellets in a 30-inch circle at 40 yards."

Although Continental shotgunners had at first snubbed Browning's "shooting machine," the engravers in the F.N. shops were fascinated by the expanse of uncluttered space offered to their chisels by the large Auto-5 receivers. The founder of the F.N. school of engraving, the master engraver Felix Funken (see the *American Rifleman*, April, 1983, page 30), used the Auto-5 to create some of his greatest masterpieces, many of which were displayed at the International Exposition of 1930 held in Liege. Funken's "Grand Deluxe" Brownings began to find favor among European firearms connoisseurs and shooters, including M. Cotty, the president of France, and King Carol of Romania.

Along with the Winchester Model 1894 rifle and Colt 1911 .45 automatic pistol, the F.N./Browning automatic shotgun has been a major contributor to the Browning reputation for longevity. For 71 years, beginning in 1903, interrupted only by two world wars, the Fabrique Nationale plant at Herstal turned out literally millions of the shotguns.

To mark the production of the 1 millionth Auto-5 made from the cessation of hostilities in Belgium in 1944 to 1961, the plant was the scene of a celebration attended by Bruce W. Browning, grandson of the inventor, and local officials. The center of interest was the millionth shotgun inscribed "1944-1961' No. 1 000 000," which now rests in the company's display room at the factory.

When the 2 millionth Auto-5 came off the line in 1970, Browning Arms Company ordered the preparation of an especially deluxe model heavily embellished with gold inlays and stocked in premium-grade walnut. This super-grade shotgun was donated to the National Shooting Sports Foundation by Browning, and was auctioned at the Shooting, Hunting and Outdoor Trade (SHOT) Show in Houston, Texas in January, 1986. The successful bidder, William H. Henkel, represented by Gary McDonald of the Old Do-minion Sports Center, Winchester, Virginia paid NSSF $50,001, which will be used by the Foundation to further its educational programs.

Advanced gun collectors have always considered the high-grade Auto-5s as prime collectibles. There were four grades illustrated in the 1931 Browning Arms Company catalog. The Standard Grade No. 1 blued with no engraving, listed at $49.75. Grade Nos. 2 ($64.75), 3 ($175.50), and 4 ($277), engraved with patterns of increasing complexity and coverage, could be had on special order. The three high-grade models were discontinued in 1940. They occasionally turn up at gun shows, often in unfired condition, and command premium prices.

When F.N. reopened the Herstal plant after World War II, a few Auto-5 guns were assembled from parts on hand, but government orders for military hardware to rearm the Belgian Army kept the plant fully occupied. To satisfy the increasing demand created by the return of thousands of American soldiers to peacetime hunting, Browning Arms Company arranged to have the Auto-5 made by Remington, which was tooled up to make its Model 11 under Browning license.

Advertised as "The American Browning," about 45,000 were made in 12-gauge, 25,000 in 16-gauge, and 20,000 in 20-gauge from 1946 to 1951. The American Browning is seldom found in other than used condition, but is still considered a desirable piece for a complete Browning collection.

In 1985, the Browning Arms Company introduced two limited-edition deluxe Auto-5s, the "Classic" and the "Gold Classic." The Classic, made in Japan with silver-gray receiver tastefully engraved and signed in Belgium, is limited to 5,000 units. Only 500 of the Gold Classic model will be made in Belgium from the few parts remaining in the sprawling F.N. warehouses and will also be engraved with game scenes and heavy gold inlays on the grayed receiver. Both models are specially stocked in high-grade Claro walnut.

In 1974, production of the Auto-5 was phased out in Belgium and begun at the Miroku plant in Japan, where the gun is still being made with no major changes from the original design, an eloquent testimonial to the genius of its inventor.

It will never be known how many automatic shotguns have been made on Browning's design. In addition to Remington, Savage also produced it as its Model 720 after Browning's patents had expired. Copies have also been produced by several firms in Europe and by Kawaguchiya in Japan. It has been estimated that more than 3 million Auto-5s are in the hands of hunters and collectors — most of them still putting meat on the table.

Buying a Used Shotgun

Rick Drury

I'd been looking for an old 870 for some time, to use as a knock-around duck gun. And the price was right, so I bought the smoothbore without giving it much thought. But it was some time before I was able to try it out. When I finally got to the skeet fields to shoot it, I noticed the gun would occasionally hang up. It had been a while since I'd used a pump gun, though, so I put the blame on myself. When it continued to malfunction, I started to suspect the gun. Careful observation, while pumping the gun slowly, showed that the gun was indeed malfunctioning.

Now, there are few shotguns that can claim the reliability of the Remington 870. You would have a hard time finding one that didn't work properly. Yet I found myself with one in my possession.

Such is the lot of the used-gun buyer. Nowhere can the phrase "buyer beware" be applied more appropriately. Yet bargains exist, too. A lifetime of shooting can be had from some used shotguns. Many are sold or traded in with just a modicum of use and wear. The buyer's problem is sorting between the two. But what to look for? And how can you get your money's worth?

First, one needs to understand value, because there are really two perspectives on the used-gun market. One is the collector's. High value is attached to originality. Any aberration from the original gun causes a loss in value. Some vintage guns that were plain Janes in their time are now going for awfully high prices because of such priorities.

But our purposes here involve utility and service. Such collector's value may be deemed artificial when discussing a gun for general use. But be aware that such "value" exists and that you may pay more than you need to if you get involved with a collector shotgun.

We must consider originality, too, but from a different perspective than the collector's. For the work that's been done may actually enhance the gun for our purposes. If it was done right. If it fits our needs. A Polychoke might make a collector gasp, but if a hunter is looking for some versatility, the addition would be just right. A cut stock will almost certainly devalue a given gun, but if that's what you have planned for your gun anyway, you will have saved some money.

A problem exists when we've no idea of the work that's been done. A single-barrel trap gun, which you might have in mind for handicap shooting, will be a big disappointment if it's found to throw Modified patterns. If you've purchased a gun that looks like new and a short time later the bluing starts wearing off because it has been poorly redone, it can make you blue.

It's probably a good idea to realize that it is a rare gun indeed that's not been tampered with in some manner. Chokes are reamed. Triggers are worked. Stocks are cut or drilled out for recoil reducers. These are alterations that do not necessarily prevent the gun from functioning properly.

A friend of mine once found a nice Super-X field model in the used-gun rack at a shop; the gun looked so clean he bought it without hesitation or inspection. When he took it out to shoot, it wouldn't work. A quick look under the forearm revealed a home-style spring located around the piston rod rather than factory issue.

Another friend brought home an old Model 11 Remington that had a Polychoke. His first round of skeet resulted in a score of 18. Since he's an AA shooter, this seemed a little strange. When we patterned the gun, we found its point-of-impact a foot right and 2 feet low at 21 yards.

So how does one discern the problems ahead of time? Where to start?

The first item to consider is whether the gun was shot to any extent. Generally speaking, clay-target guns stand a good chance of having been shot more than other guns. A week at the Grand or any large trap or skeet shoot can put a gun through more firings than a lifetime of hunting. Conversely, a hunting gun that looks like it's been through the Civil War may possess working parts with little or no wear. But we need some better indications than these generalities.

A good place to start is blue wear — not just

This article first appeared in Shotgun Sports

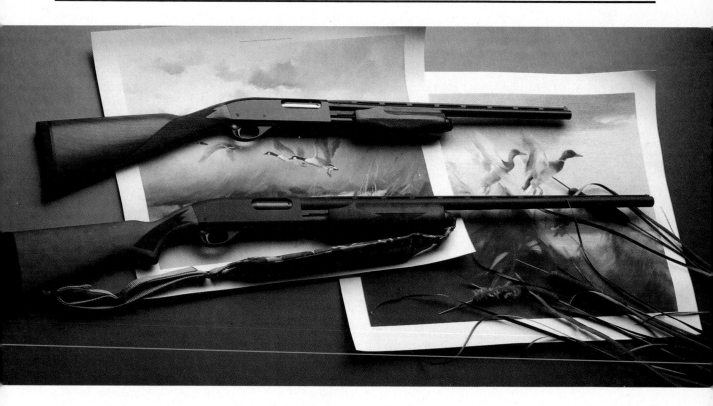

the overall looks of the gun but the wear that occurs as the gun is fired. The most obvious place to check on pumps and autoloaders is the magazine. A pump gun will reveal telltale signs of blue wear with just a bit of shooting. If the tube is worn white, you can be assured the gun has seen its share of shells. A gas gun's magazine is hidden by the fore-end, and often shooters neglect to disassemble the gun to inspect the area. But as on the pump gun, the blue wear will show. A related area on a gas gun is the inside of the barrel hanger, as it will quickly show any friction from the rings.

A point not so obvious is the bottom of the barrel extension that slides into the receiver. Shells striking upward as they are brought up by the carrier will cause blue wear and markings. With some experience, you can tell immediately if that barrel has seen much use.

Another telltale spot is where shells are slid into the magazine, at the bottom of the receiver. There will be some blue wear on the side where shells have slid forward into the magazine.

Over/unders and side-by-sides may also have some subtle wear points. My Browning skeet gun has the bluing worn off on the back of the trigger guard, as that's where I place my right hand when I open the gun.

Once I was interested in a nice Browning Pigeon Grade trap gun when I noticed that the gold on the trigger had been worn off. Now, I know my own Browning has shot well over 30,000

rounds and I've yet to wear through the gold on the trigger, so it was rather obvious that the gun had been shot a great deal.

There are many other indicators of wear also. One of the best is the face of the hammer. It will strike the firing pin with each shot, and when used a great deal, the hammer will become peened at the spot where it contacts the pin. The only tough part here is that it requires removing the stock. If it's an expensive gun you're looking at, it's worth the effort. If you find a gun with a hammer that's peened deeply, you can be assured the gun has been fired a great deal.

A bolt face will also give a good indication of use. Only a brief bit of firing will mark the face with a primer mark. On break-open guns, the back of the breech is the area to inspect. If there's been much use, it will readily show marks and dings.

Many of these parts could be replaced; however, it's unlikely that previous owners will replace them. It's your job as a potential buyer to take the time to inspect the entire gun.

Of course, care is a big variable. A gun can be shot a great deal and still be in good shape if it's been taken care of. Some shooters are conscientious. Others are downright abusive. Knowing who has owned a gun and what kind of care he takes with his firearms can tell you a lot about the gun you are interested in.

It's good to obtain knowledgeable help also, particularly if the gun's a significant investment.

Someone who is familiar with a particular type of gun, a collector or such, may be able to spot irregularities that you would pass over. I have a friend who is a Model 12 nut and is a wonder at picking out a part, or work, that isn't original.

A gunsmith can cast a knowledgeable eye on metal parts and such that we as shooters would just pass over. First, using specialized tools that the average shooter may not have access to, he'll be able to measure and test areas like trigger pull and choke dimensions. Second, he'll have access to factory specifications and be able to compare them with his findings.

Certainly, anyone involved in the used-gun trade should know how to determine whether a gun has been refinished. A reblue job, no matter how tastefully done, will show buff marks when the gun is held so that light reflects off the metal's surface. Also, watch for letters and engraving to be buffed out or stretched, and rounded-off corners. Generally, if you know what a particular factory's wood finish looks like, you'll be able to tell if a gun has been redone, since it's difficult to match most factory finishes.

Not that there is anything intrinsically wrong with a refinished gun. But there is always collector's value and, in turn, resale value to be concerned with. Further, many times the buyer will be expected to pay for the refinishing work (included in the price of the gun) when such work may actually devalue the gun! Of course, such is not the case if the gun was a mess. But often guns are refinished for practice by "new" smiths learning the trade, or by someone who thinks his old piece might be worth more if it's cleaned up. Either way, it's good to know what you're getting for the money.

Any dissertation concerning used shotguns should contain a word about choke devices. There are simply hundreds of thousands of shotguns out there with variable choke devices. Many of these devices were put on conscientiously and skillfully. Many, many others were not. The biggest fault of such devices is a bad point-of-impact. I've witnessed some as far off as 4 feet at 40 yards. The only solution is having the barrel bent to get the affair back in line.

Screw-in chokes have become popular. At first, only a few firms in the country were involved in

These photos show (from left) barrel extension with great deal of bluing worn away by shells striking upward; second barrel extension exhibiting much less wear; well-worn magazine that indicates considerable shooting; and breechface of Browning over/under that also reveals much firing.

Prospective buyer has removed barrel of nice field-version Winchester Super-X and is inspecting action and magazine.

such work, but lately gunsmiths from every direction are jumping on the bandwagon. This proliferation will surely increase the chance of sloppy work being done.

In either case, beware, and have a good gunsmith inspect the work before handing over the cold cash.

Who should you be handing that money to? Obviously, there are a host of possibilities, but a reputable dealer is probably the best bet for getting your money's worth. On the other hand, there will be fewer bargains available. Since gun trade is their business, the dealer will be aware of values and worth. Often, stores will place less value on collector's choices. Usually such dealers will stand behind their products. One of the more popular gun shops here in Pennsylvania offers a 30-day guarantee on used firearms. Almost all will at least assure that the firearm works as it should.

Individuals are often good sources of used firearms. First, there's usually less of a profit motive. A shooter might simply tire of one gun and sell so he can buy something else. In fact, the fickleness of tournament shooters, in their never-ending quest to pick off another target, can often be a source of used guns at bargain prices. Others will sell simply because they need the money, or the gun isn't being used, or a thousand other reasons.

It's good to ask for a three-day inspection. Even the most honest individual will forget work that's been done or problems that have occurred.

An inspection is particularly important when dealing through the mail, sight unseen. Most dealers will offer that courtesy, but be sure to insist on a grace period if it's not mentioned.

A third alternative in the search for a used gun is to attend gun shows. Generally they're held by gun-collecting organizations that are affiliated with the NRA (National Rifle Association). Just

about anything you could desire will be available at the larger shows. But you'll have little knowledge of a gun's history at such an affair, and though good deals are available, an intimate knowledge of the shotgun, by yourself or a friend, will be needed.

Obsolescence is another important consideration. With any gun that's out of production, there can be a problem with parts. First they become expensive, then out of production, and later nonexistent. A broken part can mean a ruined gun, a trip to the local machinist, or rummaging around the various parts companies. In any event, not a pleasant project.

Resale value should always be considered. Remember that intangible, collector's value. Well, it may be considerably less tangible when you decide to sell that collector's item. An obsolete or even discontinued gun may be seen as less valuable by a perspective buyer or dealer when you go to move the firearm.

The picture I've painted isn't exactly rosy, yet bargains are available. My 12-gauge Super-X, which I've shot trouble-free for 10 years now, was bought off a fellow for a good price. He shot 50 shells with it and then decided that he didn't like picking up empty shotshells.

Just last month I picked up a Browning A-5 12-gauge that had some rust on the outside. But I could tell it had been shot little. The internal parts were in mint condition, and I knew there was a lifetime of shooting left in the gun. The price? Only $275, a steal for a Belgium Browning.

But admittedly, I've also bought guns with cracked stocks, bent barrels, reamed-out chokes, and too may other problems to list, so if I seem overly cautious, it's born of experience.

Take time to inspect a firearm. Find out what it's been through. The bargain you're seeking is out there. Be cautious enough to insure that's what you're getting.

The Over/ Under Story

Jim Carmichel

The over/under shotgun is the most efficient wingshooting firearm yet devised. In any shotgunning game where the best wingshooters compete for the richest prizes, 19 out of 20 of them shoulder an over/under. In international-style competition, which includes Olympic skeet and Olympic trap, the choice is simple — use an over/under or you can't win. The odd competitor who doesn't use an over/under of some sort is viewed about the same way as someone on crutches trying to compete in the pole vault.

The same is true in U.S.-style trap or skeet — the top guns are over/unders. Even in England, the spiritual home of the side-by-side double, the runaway favorite in the demanding Sporting Clays tournaments is the over/under, or, as the English say, "under/over." And if you were to drop by one of the posh Spanish or South American pigeon clubs, where thousands of dollars ride on every pull of the trigger, you'd immediately notice that the high rollers are almost invariably squinting over the rib of an over/under.

THE STAR IS BORN

Despite these successes, and the obvious fact that the over/under sets the performance standard by which all other types of shotguns are measured, the remarkable fact is that for the first three-quarters of its existence, the stackpole, or "superposed," shotgun was regarded as the ugliest and most unworthy of stepchildren.

An almost equally peculiar situation is that only five (correct me if I'm wrong) hammerless over/unders have been manufactured in the United States. Of these, only one model is still in production, despite the fact that more over/unders are sold and used in this country than anywhere else!

Such legendary makers of American smoothbores as Parker, Fox, and L.C. Smith never quite caught up with what was happening and continued to make side-by-sides only, even while their market was becoming as dry and lifeless as a sackful of paper shotshell hulls.

How did the over/under go wrong at first? What caused the revolution in its favor? And what is so special about the over/under shotgun that enables it to outperform other types?

The over/under shotgun as we know it dates back no farther than the first decade of this century. Both the English and Germans claim credit for the development of the over/under, but I'm inclined to give credit to the Germans. The ritzy London firm of Boss claims to have made the first British over/under in 1909. The magnificent Boss, however, was a hand-built affair, terribly expensive, and made in such scant numbers that it did little to popularize the over/under concept.

By 1910, the Germans were offering relatively inexpensive over/unders to the gun trade but were astonished to find that there was very little interest in the new concept. To the German mind, the over/under made very good sense, and in an effort to push their product, they even sold unfinished guns to the world's gunsmiths. Thus, a gunmaker had only to fit his barrels and stocks to an inexpensive German action. A good many of the early "English" over/unders were made in this way, most of them in the gun-making city of Birmingham, England.

During the decades between the wars, German gunmakers perfected their version of what the over/under should be. The greatest of the German over/under makers were the Merkel brothers, who offered a wide range of truly splendid guns. Even their bottom-of-the-line model was an elegant boxlock, and their top model was unquestionably the greatest German shotgun. I once owned a Merkel, one of the plainer grades, which I acquired in the only smart gun trade I ever made. It produced some of the prettiest patterns I've ever seen, and when I had to sell it to

This article first appeared in Outdoor Life

Dove hunter swings light, well balanced over/under. Typically, modern "stackbarreled" guns have better balance than pumps or autos because over/unders don't need all that receiver.

pay college expenses, I went into deep mourning for nearly a year. Merkels are still being made in East Germany, but only a few trickle into this country.

BEAUTIFICATION

Despite the wonderful craftsmanship of the Merkel, it suffers from the one great failing of many over/under designs. This is the excessive depth through the action caused by stacking the barrels over the usual bottom locking arrangement. This means that the actions of many over/

unders were, or are, nearly twice as deep as those of a trim side-by-side double. In terms of strength, performance, and handling qualities, this extra depth is of little importance, but a double-barreled shotgun is nothing if it is not beautiful, and the fish-bellied over/unders hardly met the prevailing criteria of beauty. Detractors of the over/under have come up with long lists of complaints, most of them too silly to repeat, but an ugly shotgun *is* hard to live with.

The London-based firm of Boss overcame the potbellied look with an ingenious bifurcated locking system that bolted the barrels at the side rather than at the bottom. This allowed a much

Trim, graceful profile of Ruger over/under (top left) is made possible by locking system that engages barrel at receiver's sides instead of underneath. All over/unders share advantage of comparatively narrow sighting plane (top right), which many gunners find easier than sighting over side-by-side configuration. Winchester Model 101 (center), imported from Japan, is solid, dependable performer. Like most of today's shotguns, it has screw-in choke tubes that make it extremely versatile. Among world's most elegant over-unders are those made in Italy by Armi Famars. Engraved sidelock example (bottom) is a work of art.

slimmer profile through the action area and accounts for the Boss' racy profile. A similar system was adopted by Woodward, another top London gunmaker. Even today, nearly eight decades after the introduction of the Boss "under/over," the gun has yet to be equaled in terms of beauty, workmanship, and handling qualities. The Woodward, with its somewhat more massive receiver, is a favorite among pigeon shooters and was the favorite of the great bandleader Paul Whiteman, who was also something of a legend in big-money shotgunning games. In case you're wondering, the price tag for a *used* Boss or Woodward over/under usually runs more than $20,000, or you can order a new one and hope that you live long enough to take delivery.

By now, you're probably wondering where the Belgian-made Browning Superposed fits into the picture. After all, here in America it is probably the best-known over/under. More than a few good folks believe that John Moses Browning himself invented the over/under. Actually, he didn't patent his over/under design until 1926, and it was not until 1930 that the guns were manufactured. Though a relative latecomer, this great shotgun was unquestionably the crown jewel of Browning's career.

In addition to being a very nicely fitted gun offered at a reasonable price, the Browning wasn't hard to look at. Even though the locking lugs are on the underside of the barrel, Browning managed to achieve a reasonably shallow profile and gave the receiver a graceful contour. Initial sales of Browning's Superposed were slow, almost nonexistent, despite a price tag of only $75. Those were the years of the Depression.

But the Browning firm kept plugging with its over/under, and during the 1950s began a classy advertising campaign that made the Superposed the No. 1 item on every gunner's wish list. Like every other daydreaming schoolboy, I hoped to own a Browning and eventually raked up the $200 it sold for in the early 1950s. My friends thought I'd gone completely around the bend because I had bought such an expensive gun. "After all, $200 will buy an acre of farmland," my uncle pointed out. I wish he were still around today because that acre of farmland in Tennessee is worth no more than $500, but that same Superposed will fetch a quick $1,500.

One of the world's greatest over/unders, certainly one that was to have a profound effect on the course of shotgun development, was Remington's Model 32. Introduced in 1932 and discontinued with the beginning of World War II, the old Model 32 was well ahead of its time. In fact, it wasn't until decades later that competitive shooters discovered that it was one of the greatest skeet guns ever made.

Paradoxically, the Model 32 gained immortality through its imitators. The German-made Krieghoff, one of the winningest shotguns, is nothing but a copy of the original Remington design. The German maker even copied the model number. Another successful copy is the Finnish-made Valmet. It is indeed a bitter irony for American gunmakers that a U.S.-made over/under that never achieved much success during its time should become successful in the hands of foreign imitators.

The over/under bloomed fully with the arrival of good-quality but remarkably inexpensive guns made in Japan and imported by well-known companies such as Winchester. These mostly were copies of the Browning that cost only about ⅓ as much. Trap and skeet shooters who had yearned for an over/under bought the Japanese imports by the tens of thousands. Within a decade, the over/under was the gun to beat in any kind of competition. By that time, high-performance competition models made in Italy by Perazzi, Fabbri, and Beretta were also on the American market.

The skeet event of the 1984 Los Angeles Olympics was won by an American using a Remington Model 3200 over/under. Ironically, even while this gun was proving itself as one of the finest competition shotguns ever made, the Remington bosses were issuing orders to halt its production. Alas, the great Model 3200 is no more, but one hears rumors that it may be resurrected.

This leaves Ruger's trim Red Label as our only domestically produced over/under. Taking its cues from Boss and Woodward, Ruger adopted a side-locking system that gives the shotgun one of the trimmest profiles of all.

Once a hunter or competitive shotgunner tries an over/under, he usually doesn't want to use anything else. Which special quality makes the over/under so good?

There are two reasons — good balance and a single sighting plane. Nothing equals a good side-by-side double for balance and handling qualities. Many shooters, however, cannot cope with sighting over two barrels and have to opt for a gun with a single sighting plane. But with pumps and autos, the balance is bad because of the weight and length of the receiver. With the over/unders, you get both good balance (assuming it's a properly designed gun) and a single sighting plane.

Anytime I hear someone criticize the plump profile of an over/under shotgun, I remind him that we see the profile only when looking at the gun, not when shooting it. When you look at an over/under the way a shotgun is supposed to be looked at — over the receiver and down the rib — nothing looks better.

PART THREE

HANDGUNS

THE GLOCK 17 PISTOL

Pete Dickey

Designer Gaston Glock (left) easily explained his pistol to NRA staffers but was mystified by press reports on its construction. Photo on facing page provides close view of this 9mm Parabellum auto. Its magazine holds up to 17 rounds.

There has been a lot in the news media lately on the Glock 17 pistol. Most of it, as might be expected of the antigun press' coverage of any firearms, is pure hogwash.

Without going into details that would lead to irritation at best and apoplexy at worst, it should be said that:

A. The Glock 17 is *not* an "all plastic gun" — it is, in fact and weight, about 83% steel.

B. The Glock 17 does *not* "pass through metal detectors undetected." Nineteen ounces of steel plus about 4 ounces of lead, if a pistol and a full magazine are considered, should trip any metal detector — provided the unit is plugged in.

C. The Glock 17 is *not* "invisible when passed through an X-ray screen." It looks like what it is — again, provided the machine is plugged in, and assuming the viewer knows his or her business and is attending to it.

D. Given A, B, and C, we are really baffled by this one: "The Libyans are said to be trying covert methods to obtain these weapons." Why is the Glock 17 better for Libyans than other pistols? Since the Glock 17s are on the world market, why

the "covert methods?" Last, but not least, who started the rumor and why — other than to make a catchy story? We don't know, and nobody in the U.S. or Austrian government seems to either. Nor does Gaston Glock, the pistol's designer, manufacturer, and chairman of the board of Glock G.m.b.H. of Deutsch-Wagram, Austria.

Glock visited the NRA (National Rifle Association) during the height of the anti-gun press idiocy on his pistol. Together with Wolfgang Riedl, his marketing manager, and Karl Walter,

This article first appeared in American Rifleman

the U.S. importer, he could throw no light on the Libyan tall-tale source or make anything of the press reports on his "detection-proof pistol." So much for the rumors.

Glock, however, has many facts at his disposal, and we took advantage of his visit to get some first-hand information on what turns out to be a very interesting — but not supernatural — pistol.

His answer to our first question regarding his gun background was a real eye opener.

"In 1980, I didn't know the difference between a pistol and a revolver," Glock told us. Things progressed from there.

In 1963, Glock, a mechanical engineer with a background in synthetic materials, formed a company to produce furniture and hardware. The company eventually made combat knives, bayonets, and entrenching shovels for the Austrian military that are still in the line and available in the U.S.

In 1980, Glock saw a copy of the Austrian military's pistol requirement form, but thought little of it. Essentially, it called for 9mm Parabellum chambering, a large magazine capacity, light weight, and durability.

Then, however, his 15-man factory was approached by a foreign pistol maker who was interested in the Austrian requirement but wanted an Austrian manufacturing or assembly site — preferably Glock G.m.b.H. No deal was made, but Glock began to take matters seriously. When

two more foreign pistol makers approached him with similar proposals, he made his decision.

He studied military pistols, test procedures, and requirements from all over the world and, with two of his employees, turned out a working prototype in exactly six months. It didn't suit him in every particular, so he made a second prototype in two months.

That prototype is, in every detail, the same as the pistol now used by the Austrian Army and issued in a quantity of 28,000 pieces.

If there is a similar success story around that involves a man innocent of even basic gun knowledge who designs a totally new gun, we are unaware of it.

At any rate, the deed was done and a major European nation is using the Glock. Is any other nation using it? Yes, Glock told us, Norway has accepted the Glock 17, and in three years the Norwegian Army will be totally armed with it. This is of more than passing interest for, while Austria is not, Norway *is* a NATO member and the Glock is, therefore, a NATO standard pistol. In addition, we were told, many military and police units are considering the Glock 17, and some have purchased it in more than trial quantities. Not the Libyans, however.

"Does the factory still have only 15 workers?" we naively asked. Glock smiled and told us that, no, there are about 40 involved in the pristol project, but one man controls the computer panel

Glock pistol, using modern materials and original design, is not only innovative but simple, durable, light, and accurate. Upper photo shows gun's external appearance. Lower photo is X-ray view of fully loaded Glock, provided by Metallurgical Engineers of Atlanta (FAA Certification No. 701-31).

that, in turn, controls all the ultra-modern, fully computerized machinery.

Here, Glock feels, is one of his great assets. He has visited many arms factories in the last few years and confesses to amazement at the great amount of "obsolete equipment mixed in with some modern equipment" found in virtually all of them. Except his.

What use does Glock G.m.b.H. make of sub-suppliers, we asked, and again got a surprising answer and another reason that Glock thinks he has an advantage over the competition.

"We make absolutely everything ourselves except the springs and raw material. The steel slide material is received in profiled bar form; we mill it to shape. The polymer is received to our patented specifications; we mold it and so forth. We also do the finishing, of course, and use three hardening processes for the slide and barrel. The final hardening produces the patented 70 Rockwell cone finish that is harder than a file."

A two-part question was what Glock's next firearm project would be and if he visualized an all-plastic gun.

To this, he said his next gun, if any, would be another one tailored to Austrian military requirements. He feels — and he should know — that plastics or polymers have not reached the stage where they could be used alone to produce a military-acceptable firearm.

As Glock was, with commendable speed, fully disassembling one of his pistols with only a small punch (a nail would have done nicely), we gave him a final question.

"How does the U.S.-imported Glock differ from your Austrian and Norwegian service pistols?"

We were told that the only differences, aside from markings, are that to the U.S. pistol was added the molded-in serial number plate in the polymer frame, and the rear sight that is click-adjustable for elevation. Period.

Other questions remained, but in the absence of Messrs. Glock, Riedl and Walter, these were best answered by a normal examination of and test session with the intriguing new pistol.

The Glock 17 is a locked-breech, hammerless pistol that is sometimes called a "double-action." The term is misleading, and while Glock's preferred term "Safe-Action" is not self-explanatory, Glock's terminology and parts illustration are used here to explain matters.

When the slide (1) is retracted and released, the firing pin (5) is partially held back ("half cocked") by the cruciform sear plate on the rear of the trigger bar (26). This does several things: it keeps the firing pin from primer contact: it allows the passive firing pin safety plunger (9) to block pin movement; and it shortens the necessary final movement or cocking of the firing pin by trigger depression when discharge is wanted.

The pistol always works that way whether the slide is retracted manually to load the chamber for the first shot or by recoil for subsequent shots. Thus, there is no difference in the trigger pulls for the first vs. subsequent shots. Such different pulls are necessary with conventional double-actions that really operate as "long-pull" double-actions for only the first shot and "short-pull," self-cocking single-actions thereafter.

The sample Glock 17 that we tested had a smooth, constant, medium-length pull of about 6 pounds, but lighter or heavier pulls can be had by substituting the appropriate trigger springs that are available from the importer.

The trigger itself contains the pistol's only manual "safety." This is a pivoted lever that protrudes through the face of the trigger and extends through the trigger's body to emerge again at its top rear where it contacts the frame. When the trigger is depressed by the shooter in normal fashion, the safety lever pivots and its rear portion moves up into the trigger body and out of contact with the frame. The trigger is then free to move backward and effect discharge.

The depression of the safety lever requires no effort as its return spring is very weak. The Safe-Action name is derived from it, and the main advantages of the system are simplicity and insurance against firing if the loaded pistol is accidentally dropped. If, as in a combat situation, the decision is made to carry the Glock with the chamber loaded, the soldier need not question whether a conventional thumb safety is "on" or "off." The pistol will fire when the trigger is pulled, for the safety lever will automatically be pivoted out of engagement with the frame.

Obviously, this is unusual, and Glock's manual stresses the point that the pistol ordinarily should be carried "empty, with the trigger rearward except when you intend to shoot." It is impossible to carry or set the pistol so that the trigger remains back when a round is chambered.

The trigger's safety lever, then, can be said to prevent the trigger from being depressed until purposely depressed by the trigger finger. Its presence and the absence of any other manual safety require that particular attention be paid to familiarization with the Glock 17 by a prospective user, particularly one used to a more conventional pistol.

Aside from the trigger, trigger safety, and "half-cock" firing pin systems, the mechanics of the Glock 17 are fairly straightforward.

Takedown is accomplished by first pressing the magazine catch (19) located on the left of the frame behind the trigger; the magazine will drop of its own weight from the beveled magazine well.

The slide is then opened and closed to make sure the chamber is empty. Then the trigger is pulled and will remain in its rearward position; the slide is opened ¼ inch or so, and the locking slide (21) above the trigger is pulled down.

With the locking slide depressed, the slide unit is slid off the front of the frame. Now the recoil spring and tube (3 & 4) can be removed, followed by the SIG-type barrel (2) with its squared chamber area that forms the massive locking lug. The barrel's underside has an open cam (also reminiscent of SIG) that engages the steel locking block (22) molded into the polymer receiver.

That completes the field-stripping of the Glock 17, and further disassembly, though possible with only a nail, is not recommended by the factory.

The polymer receiver, when stripped of the slide unit, reveals the working trigger/sear parts, the locking block, and the slide stop lever (27). All are of steel as are the four molded-in slide "rails" that measure about .4-inch in length and are located in pairs at the rear of the receiver and

Glock 17 Parts Nomenclature

P = Polymer
S = All steel
P/S = Polymer with steel inserts or appendages

S1. Slide
S2. Barrel
S3. Recoil spring
P4. Recoil spring tube
S5. Firing pin
P6. Spacer sleeve
S7. Firing pin spring
P8. Spring cups (2)
S9. Firing pin safety
S10. Firing pin safety spring
S11. Extractor
S12. Extractor plunger
S13. Extractor plunger spring
S14. Spring-loaded bearing
P/S15. Slide cover plate
P/S16. Rear sight
P/S17. Receiver
S18. Magazine catch spring
P19. Magazine catch
S20. Locking slide spring
S21. Locking slide
S22. Locking block
P/S23. Mechanism housing with ejector
S24. Connector
S25. Trigger spring
P/S26. Trigger with trigger bar
S27. Slide stop lever
S28. Trigger pin
P29. Mechanism housing pin
P30. Follower
S31. Magazine spring
P32. Magazine floor plate
P/S33. Magazine tube
P34. Front sight

ACCURACY RESULTS
Five Consecutive 5-Shot Groups At 25 Yds.
Fired From Sandbags

9mm Luger/Parabellum Cartridge	Vel.@15' (f.p.s.)	Smallest (ins.)	Largest (ins.)	Average (ins.)	25 Shot Composite (ins.)
CCI Lawman 3610 100-gr. JHP	1255 Avg. 12 Sd	1.63	3.31	2.35	3.76
Federal No. 9AP 123-gr. FMJ	1125 Avg. 16 Sd	2.45	4.25	3.27	4.77
PMC 9A 115-gr. FMJ	1145 Avg. 21 Sd	3.25	5.41	4.49	6.89
Average Extreme Spread				3.37	

GLOCK 17 PISTOL

Specifications

Manufacturer: Glock G.m.b.H., Hausfeldstrasse 17, A-2232 Deutsch-Wagram, Austria
Importer: Glock Inc., 5000 Highlands Parkway, Suite 190, Smyrna, GA 30080
Mechanism Type: recoil operated, semi-automatic
Caliber: 9mm Parabellum
Overall Length: 8 inches
Barrel Length: 4⁵⁄₁₆ inches
Weight: 24 ounces
Height: 5¼ inches
Width: 1³⁄₁₆ inches

Magazine Capacity: 17 rounds with contents indicator
Trigger: "half-cocking" double-action only, 6 pounds
Sights: Patridge-style front sight with white dot, rear notch with white outline, click adjustable for elevation, drift for windage
Stock: integral with receiver
Accessories: air-tight plastic carrying case, cleaning rod with brush, extra magazine, loader, and instruction booklet

FIELD STRIPPING GLOCK PISTOL

Glock 17 is easily field stripped by first removing magazine and clearing chamber. Then slide is retracted about ¼" and locking slide is depressed (Fig. 1). Slide unit can then be slid forward from frame (Fig. 2). Recoil spring and its tube are then removed (Fig. 3), followed by barrel (Fig. 4). Steel rails (arrows) are now revealed, together with many other steel parts that account for both the locking and the Fast-Action firing mechanism (Fig. 5).

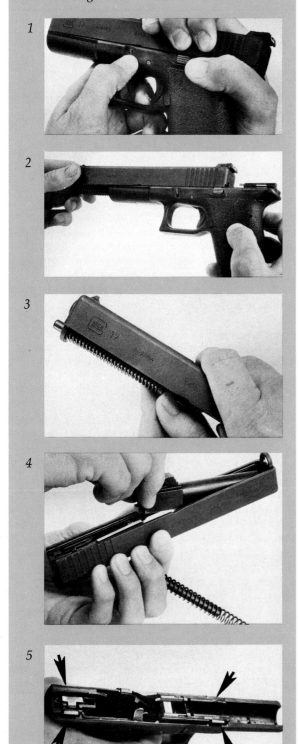

above and forward of the two-hand-hold trigger guard.

The broad, flat-topped front sight is .17-inch wide with a white dot insert; the flat-topped rear sight, with an inserted white bracket, has a .125-inch-wide square notch, and the resultant 7-inch-radius sight picture is excellent. Nevertheless, some doubts were voiced by shooters prior to accuracy and function testing.

What was expected was a certain military crudity in handling qualities and heavy recoil from the polymer-framed pistol's light weight. What was discovered was that the Glock is in fact an unusually pleasant pistol to fire.

The Glock's firing pin is worthy of mention, as it is rectangular in cross section, with a convex nose. The fired primer takes on the raised rectangular impression of the slot-like firing pin hole in the breech, with a firing pin indent in the center. It looks odd but, we found, works well.

The 17-round steel-lined magazine is loaded with the aid of a plastic loading tool packed with the gun. It is a sleeve that telescopes the magazine. A wedge inside its top presses downward on the magazine follower, allowing easier insertion of ammunition. Its action takes a little familiarization, but once it is mastered, a magazine can be loaded in short order, with no sore fingers.

Firers anticipated that the Glock's trigger arrangement, requiring half-cocking for every shot, would result in a springy, hard-to-manage pull. This was not the case. While it is hardly a competition trigger, the Glock's 6-pound pull is an easy one to master, and has the advantage over conventional double-actions of a consistent pull on all shots.

The pistol's pointing qualities make it hard to believe that its designer is a man of limited firearms experience. The grip is very well designed, from the standpoints of recoil absorption and instinctive pointing. Both ham-fisted and diminutive firers found the trigger and slide release easy to reach. Rapid-fire, either one- or two-handed, was facilitated by the pistol's limited muzzle jump. In short, the Glock 17 handles quite as well as any large-capacity 9mm tested in recent memory.

Over 300 rounds of mixed ammunition have been fired from our sample Glock 17. There were no malfunctions of any kind, and representative accuracy figures are shown in the table.

That sounds good, and it is, but, according to Glock, the Austrian military has repeatedly fired five-hour test sessions of 10,000 rounds each "successfully," that is to say, without a single misfire or malfunction.

Austrian soldiers are, apparently, a quick-shooting, hardy lot, and the Glock 17 should suit them well.

Is There an Ideal Hunting Handgun? . . .

Bob Milek and Ross Seyfried

Two top experts thoroughly disagree about what constitutes a legitimate hunting handgun. Bob Milek states his case at left; Ross Seyfried holds forth at right. You be the judge!

Some 42 years ago, I poked six rounds of .38 Special ammo into the chambers of my dad's Colt Officer's Model Match revolver, hitched up my pants, and wandered into a sage-rimmed dry gulch in search of a mess of cottontails for our evening meal. Even back then, a handgun was a tool for me—a handsome, beautifully balanced tool to be sure—but still a tool with which I plied my favorite sport—hunting.

In the intervening years I've devoted a good share of my time to the handgun, cartridges to shoot in it, and to the sport of handgun hunting. Unlike many handgunners, I've been quick to accept the single-shot (break-open, bolt-action, and falling block) pistols, handgun scopes, and modern high-velocity cartridges that have brought the sport of handgun-hunting out of the dark ages. I'm especially proud of the fact that I've been directly involved in the development of some of these products designed to greatly extend the range and efficiency of the handgun hunter.

After all these years, I still look at a handgun as a tool. I have a job for it to do, and any handgun that can't do the job as efficiently as possible has no place in my gun cabinet. As I see it, a handgun is a sporting firearm to be used for hunting. I'm not now and have never been into competitive handgunning, be it paper punching, metallic silhouette shooting, combat competition, or benchrest matches. I have no argument with those who derive hours of satisfaction from these sports, but I'm a hunter. I'd hazard a guess that

This article first appeared in Guns & Ammo

of the thousands of rounds I fire through handguns each year, less than half are fired in practice, load development, testing, and sighting in. All the rest are fired at live game in the field. This isn't difficult to believe when you realize that I go through several 5-quart ice cream buckets full of .223 Remington and 6mm/.223 handloads each spring just shooting varmints.

Obviously, my definition of a handgun is far different from that of those whose interests lie in other areas of the handgunning sports. Nevertheless, because of the diverse shooting conditions encountered in the field and the variation in game size and stamina, I have some guns in common with just about every pistol shooter regardless of interest.

Any attempt on my part to define what a handgun is to me in mechanical terms would be nothing more than a repeat of the hundreds of articles I've written on pistols and handgunning over the past quarter century. Instead, I'll broach the subject from a philosophical bent because the handguns I use and how I use them are a manifestation of a basic handgunning philosophy I've developed over the years and one to which I rigidly adhere.

Right now is probably a good time to clear up one point. Contrary to some opinions, handgun cartridges do not fall into any special niche when it comes to how they perform on game. Ballistically, handgun cartridges must stand comparison with rifle cartridges. The fact that a certain handgun has a bore the size of a sewer pipe and shoots a cartridge that belches fire, roars like a jet engine, and kicks like a loco mule has nothing

(continued on page 72)

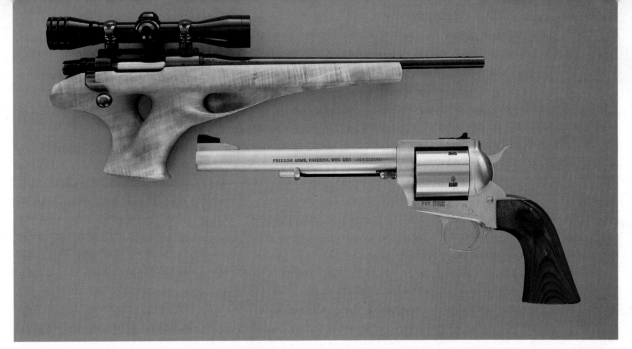

Seyfried maintains that anyone hunting with long-barreled, scoped pistols like custom-stocked Remington XP-100 bolt-action shown here is really using "short rifle" and should not call himself a handgun hunter. Milek believes fixed open sights have no place in hunting, and pursuit of dangerous game with such sights, even on handguns as potent as .454 Casull revolver in this picture, is just exercising ego.

The very first firearms were by strict definition handguns because they didn't have stocks that touched the shooter's shoulder. I'm talking about bamboo tubes full of black powder and rocks. Some evolution has taken place since then, and possibly some digression, because today every firearm that doesn't touch the shooter's shoulder is called a handgun. In my opinion, that isn't how we should define handgun. There is really only one basic reason for a handgun, *portability*. It is this quality that makes a handgun. The handgun is in fact fired without touching the shooter's shoulder, but it is how you carry it that makes it a real handgun. The holster is the primary means, but pockets, purses, garters, briefcases, and saddlebags also fill the bill. There are too many ways of carrying a handgun to list, but there are two "carries" that make a firearm *not* a handgun: both hands or a sling. That is, if you need a sling or both hands to carry your gun, it ain't a handgun. But take heart, it still might qualify for "short rifle" status.

Handguns were originally designed primarily for personal defense. Personal-defense handguns rarely get caught up in the confusion. Personal defense is a real world that doesn't allow much room for misdefinition of the tools. The very properties of portability, concealment, and fast accuracy pretty well keep that closet clean. Someone wandering around with a 5-pound, scoped single-shot slung around his neck would look a little silly in the FBI academy, and if he really needed to defend his person with his "handgun," he would probably die for his mistake.

There are two environments where the handgun has almost lost its identity. They are the grand sports of hunting and silhouette shooting. The I.H.M.S.A. (International Handgun Metallic Silhouette Association) has allowed the word handgun to be stretched into a grotesque creature. Winning in their unlimited world is done with a short rifle, not a handgun. Heavy single-shots chambered for rifle cartridges are the order of the day. These fellows need to take a look at their title. There are some glowing exceptions in their sport. These are the classes that require the shooters to use revolvers or shoot standing. Now the fellow who stands on his hind feet and hammers steel rams at 200 meters or more with a production revolver is damn sure a handgunner—the kind of fellow that Elmer Keith would take his hat off to!

Handgunning hunting suffers from an even more advanced case of misdefinition than the competition sports do. There are no rules in hunting to define handgun; the accepted criterion seems to be only that the gun isn't fired from the shoulder. The handgun hunter can use almost any contraption he wants and still call himself a handgunner. Single-shots of all descriptions, bolt-actions chambered for rifle cartridges included, still qualify as handguns. Add a sling and a high-powered scope, and you will be even more at home in this crowd. I really have only two problems with their misuse of the word handgun: first, it takes the honor away from the hunter who really is a handgun hunter, and second, many lose sight of just how underpowered their short rifles really are.

I am a handgun hunter and have the greatest

(continued on page 73)

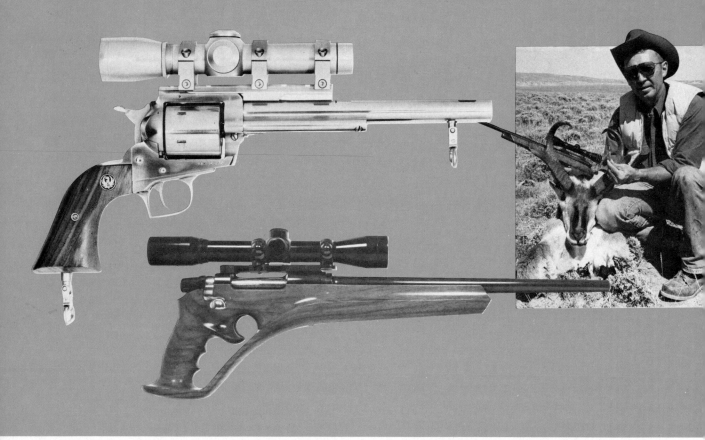

Milek favors scoped, custom handguns such as Mag-na-port Stalker (top left) and Remington XP-100 (right) for hunting antelope. Scoped, custom XP-100 Remington chambered for the 6mm/.223 cartridge (bottom) is favorite of Milek's for hunting pronghorns. He believes such guns have brought handgun hunting out of the dark ages.

MILEK'S ANSWER *(continued)*

whatsoever to do with its pertormance on game. It may be the hardest kickin', most powerful handgun available, but when it' s compared ballistically with any of our better hunting rifle cartridges, it'll come off a poor second. We can't rewrite the book on ballistics to support our prejudices concerning a specific handgun or cartridge.

I've been in the hunting game for a long time—over 40 years, both professionally and for my own pleasure—so I've had ample opportunity to observe firsthand how both rifle and handgun cartridges perform on a wide variety of game. It's my opinion that no handgun cartridge presently available, commercial or wildcat, is adequate for taking big, dangerous game. Sure, it's been done and will be again. But what has this to do with anything? That's like saying that because someone navigated the Grand Canyon of the Colorado in a canoe, canoes are the proper craft for floating the Grand Canyon. Canoeing the Grand Canyon and handgunning big, dangerous game are nothing but tricks designed to bolster the ego of the participants.

There's no room for tricks in the hunting field,

and any time you go after big, dangerous—game that professional hunters recommend be taken with rifle cartridges like the .338 Winchester Magnum, .375 H&H Magnum, or .458 Winchester Magnum—you're playing a game that's grossly unfair to your quarry and damaging to the credibility of the entire sport of handgun hunting. The same is true when you choose to use a handgun/cartridge combination that's inadequate for the job you have for it or stretch the shooting distance at game beyond that at which handguns can be effective. There isn't a handgun cartridge arond that will consistently produce one-shot kills on elk and moose at over 100 yards. And those that do the job at 100 yards are not only few and far between, but are for the most part available only in specialty pistols.

You must understand that my concern is for the game animals themselves and for the image we project to the world about the sport of handgun hunting. I couldn't care less about a handgunner who goes after dangerous game and gets himself mauled. That's his fault. The same is true of the guy who fails to use enough gun, then moans about his misfortune at losing a trophy

(continued on page 74)

For big, dangerous game, Seyfried likes wildcat magnum handguns with open sights. He took buffalo with Linebaugh .45. Linebaugh custom Seville pictured here, one of his favorite handguns, is chambered for .45 Colt Magnum. He is currently experimenting with a more potent cartridge, .51 Magnum.

SEYFRIED'S ANSWER *(continued)*

admiration for those who pursue the sport in its true sense. Handgun hunting by my definition can be described with two words: skill and difficulty. Hunting with a real handgun (one that you can carry in a holster, that has open sights, that you can hit quickly with, and that has a barrel of less than 10 inches) is a tremendous challenge. This hunter, if he is successful, has to be a great shot, a superior woodsman, and, most of all, be prepared to go home without ever firing a shot. Because his tools are mechanically inferior to a rifle, he must refuse shots in poor light or ones that are too far away for his skill level.

Elmer Keith's skill level ranged to several hundred yards with his pet 4-inch .44 Magnum. Most ordinary men and women will have to quit shooting at less than 100 yards. The essence of handgun hunting comes not from how far you shoot, but from how well you hunt. The handgun hunter who can consistently take bucks through the ribs after he has stalked to within 75 yards, using a real handgun, is not only an awesome field marksman, but a skilled hunter as well.

The fellow who zaps a buck 200 yards away with a scoped bolt-action chambered for a rifle cartridge has only killed a deer by comparison. The other problem with our fellow and his short rifle shooting big game at long range is that he often forgets that he still has an arm with relatively low power.

If I were to write a story within these pages praising the virtue of the .30/30 with its 150-grain bullets traveling 2,300 fps as a long-range car-

tridge, I would be laughed right out of print. The short riflemen with their 7mm BR, .357 Herrett or .35 Remingtons are firing bullets of 150-200 grains at velocities just over 2,000 fps and *thinking* they have long-range arms because their bullets are going faster than a normal .44 Magnum. The real truth is that they have arms that are extremely marginal performers on animals over 100 pounds. If they stretch the range too far, wounding is a very likely possibility. These arms are so shootable (easy to hit with) with their optical sights that if the shooter is given time to wiggle into some benchrest position, he can easily hit game at ranges well beyond the cartridge's ability to make clean kills.

The addition of an optical sight to an ordinary handgun more or less excludes it from my definition of a handgun where hunting is concerned. The scope makes the gun cumbersome to carry and makes it much easier to hit with. I'm the last one to say anything against any aid for precise shot placement, especially on game animals, but when we talk about handgun hunting in its real sense, shooting skill must take the place over mechanical devices. Hunting skills must put the hunter within the range that he can handle with iron sights.

There is one exception: hunters whose failing eyesight makes the use of iron sights impossible. For them the scope is the only alternative. The hunting handgun fitted with a scope of 2X or less reopens the door to handgun hunting for them.

Now that I have said what I believe a handgun
(continued on page 75)

MILEK'S ANSWER (continued)

buck or bull. What's important is that such hunters have deviated from what should be the unspoken creed of any sportsman worth his salt—to dispatch his quarry with a single, well-placed shot. To knowingly do anything else is to admit a total disregard for the animal being hunted.

Not only must the handgun/cartridge combination you use be powerful enough to handle the job you have for it, but is must be accurate enough that you can place your shots with confidence to effect quick, one-shot kills at the maximum anticipated range. It never ceases to amaze me that the same hunter who goes into convulsions of frustration if his deer rifle won't hold five shots in under 2 inches at 100 yards will dance with glee when the handgun he intends to hunt deer with will barely hold five shots in an 8-inch circle at 100 yards. With few exceptions, I set the same standards for my handguns as I do for my hunting rifles. If the handgun won't perform, I get rid of it. The only exceptions I make to this rule are for close-range hunting. A handgun to be used on small game or varmints out to say 50 yards must be capable of 3-inch groups at this distance, and I'll settle for 4-inch, 50-yard groups from a revolver I intend to use on deer-size game at this range. But when I get

out there to 100 yards, things must tighten up. I won't hunt small varmints at long range with a pistol that won't hold five shots in 1-inch or under at 100 yards, and all of my big-game hunting handguns must produce 100-yard, five-shot groups no larger than 2 inches.

Make no mistake about it, there are a lot of pistols and cartridges out there that are capable of delivering the accuracy I require of them. However, whether they will or not is often dependent upon the sights being used. As far as I'm concerned, fixed open sights have no place in the hunting field. I'll use fully adjustable open sights for small game and varmint shooting out to about 50 yards, and good open sights are okay for hunting big game out to 75 or so. But, once you pass that 75-yard mark, nothing does the job like a good long-eye-relief pistol scope.

I'll be the first to admit that a pistol scope adds bulk to a handgun and detracts from the traditional good looks of some pistols. That's why I seldom scope any handgun I intend to use at short range. However, scopes have played a major role in extending the effective range of the handgun, and for any long-range work, their advantages considerably outweigh the disadvantages. A scope does many things. First, it mag-

(continued on page 76)

Custom bolt-action (top) is a wild looker with striped stock. T/C pistol (center) features interchangeable barrels. Remington XP-100 (bottom) has become so popular with handgun hunters that aftermarket industry has sprung up to supply custom stocks, scopes, and rings.

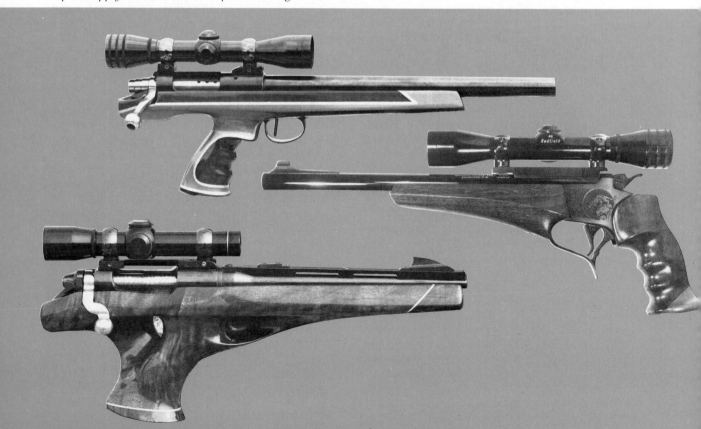

SEYFRIED'S ANSWER *(continued)*

isn't, it's time to look at what a handgun is. The "teaching" handguns, those you should learn to master handgunning with, are the .22s or the .38 Specials. An airgun that is more or less shaped like a normal firearm is even better due to its lack of noise or recoil. Revolvers or autos with barrels in the 4- to 6-inch range are just right. Handguns used for personal defense more or less dictate what they need to be in size and shape; the actual cartridges are subject to continual debate. Anything that goes bang will be better than fingernails. I pick the .45 auto.

Handguns used for hunting the most confusing. In most cases, I prefer to use a handgun in the hunting fields as a tool of opportunity. (That, by the way, was the original intent of a handgun: a tool that was there when you needed it.) Generally speaking, the hunting handgun is at its best when it is carried as a companion to a rifle. A handgun with a sensible size and shape, carried in either a belt or a shoulder holster, can be used to take advantage of shots within the capability of both the cartridge and the shooter. A hunter can take advantage of a good stalk, and other conditions suited to the handgun, while still having his rifle to take the extraordinary trophy at long range or just bring home the bacon when the handgun fails.

Handguns suited ot hunting either as a companion to a rifle or as a primary arm are more or less the same. The selection of cartridges and loads is relatively limited. The .41 Magnum is a little light for anything bigger than small whitetails, but loaded with long, heavy bullets it can be effective. The .44 Magnum will probably always be the king of hunting handguns for all but the most specialized hunters and conditions. The .44 is all right with the standard 240-grain Keith bullets, but heavier bullets make it a lot more gun. In the S&W Model 29, I like to use a 275-grain bullet of the Keith shape. In Ruger's Redhawk or Blackhawk and various single-shots, the truncated cone bullets weighing 300 to 320 grains offered by SSK and others are the very best hunting loads available for the .44. The .45 Long Colt using 310-grain Keith bullets in the strong Rugers or single-shots is even better than the .44.

Specialized revolvers chambered for the superpowered .45s—Linebaugh conversions and the Casull revolvers—using cast bullets weighing in the 350-grain range are the finest big-game handguns in the world today. The last class, the super heavyweights, belongs to the .510 Linebaugh and the new .475 revolver that is being made for me. These are tools of the specialists, designed for hunting dangerous game or for taking the biggest game on earth. They launch bullets over 400 grains at velocities in excess of 1,200 fps. While they are still underpowered compared to really heavy rifles like the .458 Winchester, they begin to stand shoulder to shoulder with ordinary rifles. Their huge bullets with deep penetration have the power to actually break down animals over 1,000 pounds. There is a drawback, though:

(continued on page 76)

At top of Seyfried's list of all-time favorite handguns is S&W .44 Magnum (left). He feels that .44 Magnum is king of hunting handguns and is suitable for all but a few specialized hunting situations. Linebaugh Ruger Bisley conversion (center) is .51 Magnum. While Seyfried espouses cartridges of .44 caliber and above, he has confessed to secret love affair with 2½-inch S&W .357 Mag.

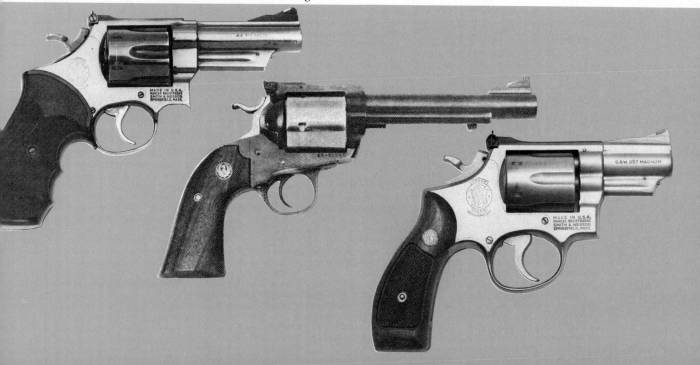

MILEK'S ANSWER *(continued)*

nifies your target, so you see better. Second, it places the reticle and the target on the same plane so both the target and aiming device are in sharp focus. Third, the reticle covers a small area, even at 200 yards—much smaller than that covered by the front blade of an open sight, thus a lot of shooter error is eliminated. Granted, handgun scope technology is in its infancy as compared to that of rifle scopes, but the long-eye-relief pistol scope of today is many times superior to what it was just a few short years ago, and if you buy the best, you can count on it to perform.

It should go without saying that a hunting handgun must be strong and reliable, capable of taking extremes of heat and cold, dust and moisture in stride and absorbing the inevitable knocks of the field without breaking down. This is a pretty tall order, but I find that most of the quality handguns on the market today will pass this test without difficulty.

This brings us to the subject of which handguns, at least handgun designs, are best for hunting? If accuracy and long-range capability were the only prerequisites, there's no doubt that the specialty single-shot pistols like the Thompson/Center Contender and Remington XP-100 would win hands down. However, these aren't always the foremost considerations. There are times when quick-handling characteristics or the need for fast second and third shots are equally important. In some situations there's nothing better than an open-sighted big-bore revolver, while other times a good semi-automatic .22 Long Rifle is just what the doctor ordered. I guess what I'm getting at is that there's no way that anyone can say this handgun or that one is best for hunting. The gun, the cartridge, and the sighting equipment must be matched to the particular hunting situation. Obviously, this opens the whole field of available handguns to the hunter, making him the most versatile user of handguns in the world.

What all of this boils down to is that to me a handgun is a sporting firearm, one to be used in the hunting field. Handgun hunting is a way of life for me, and it's gratifying to know that it's one of our fastest-growing shooting sports. When used properly, a handgun is an efficient hunting firearm. I have no time for those who, either through misuse of the handgun or their failure to confine its use on game to within its limitations, bring about criticism of the handgun hunting sport from inside and outside the ranks of American sportsmen. There are few handguns that the hunter can't find some bona fide use for in the field. The trick is to choose the right handgun and cartridge for the job, then to be wise enough to know when you're asking it for performance beyond its capabilities or yours.

SEYFRIED'S ANSWER *(continued)*

recoil reaches levels that only a few men will ever master, and the hunter who hunts big game must master these guns if he is going to use them in the field.

The actual guns used as launching pads for these hunting handgun loads leave the hunter with a wide choice. Revolvers or single-shots with iron sights both qualify. Generally, barrels should be under 8 inches long if the hunter is going to carry them effectively. I won't try to carry a handgun with more than 6 inches of barrel. While the longer barrels give higher velocities and longer sight radius, they are so cumbersome to carry that I lose interest in them quickly. Yes, the single-shots are fine hunting handguns. Like the single-shot rifles, they are an added challenge, and an added incentive to place the first shot with absolute percision. A T.C. Contender with a 6- or 8-inch barrel chambered for the .45 Colt or even the .444 Marlin round is an exceptional hunting piece.

I suppose this piece won't be complete until I tell you what my all-time favorite handguns really are. I confess that for some reason I have very little use for .22 Long Rifle handguns, so I will leave them off. The 4-inch-barreled S&W .44 Mag. has been such a part of my life that it may always head the list. My secret love affair with a 2½-inch-barreled Model 19 S&W .357 Mag. isn't revealed too often. This little fellow is delightful to carry, extremely fast, and has taken game at over 100 yards for me. I like it best because it is the one gun that really taught me to shoot. I spent years learning to master it, both single-and double-action. The short barrel is absolutely unforgiving, but if you do everything right, its accuracy is awesome. For personal defense, if I can't have a shotgun to hide behind, I want one of my supertuned Colt .45 auto or my old Colt Commander polished by Armand Swenson. John Linebaugh's .45 Colt made on an extinct El Dorado frame is the finest hunting handgun I have ever used. Its 5½-inch barrel balances perfectly and delivers a 345-grain bullet at just under 1,500 fps. With all of its horse-power, this gun hits better for me than any handgun I have ever fired. It was this gun that went with me into the thorns of Africa after a wounded Cape buffalo. Walking out unscratched made me friends with this gun for life. When I get one just like it made on a Ruger Bisley, it will be my favorite shooter.

I am sure some of what I have said about handguns has invited your hate mail, but be sure you don't misunderstand. Just because what you are shooting isn't a handgun by my definition doesn't mean I am not on your side. We are all shooters together. Just be sure that if you call yourself a handgunner, you deserve the honor.

Choosing a Used Handgun

W. E. Sprague

THERE is a certain delight in bringing home a brand-new gun. There is an indescribable pleasure in removing the new treasure from its factory box and wrapper, wiping it down with a fresh, clean cloth, and viewing its bright, unblemished luster.

But that pleasure has a definite cost. A new gun, especially if bought at full retail price, loses much of its value with the first shot fired.

In many ways, a used gun is the easiest to buy with confidence. An individual seller or responsible dealer should be willing to make a conditional sale, giving you the chance to test-fire the gun for reliable function.

Furthermore, in buying a used gun, you won't be paying a new-gun premium; you should be able to sell it for nearly as much as or maybe even more than you paid for it.

Despite the financial advantages, many shooters are reluctant to buy a used pistol. Everyone has heard horror stories about what seemed to be bargains that turned into repair-bill nightmares. But with a few basic inspection techniques, you can avoid buying that "lemon" when shopping for a used sidearm.

The first and most obvious thing to check is the finish. Is the plating or bluing original, and is it badly worn, scratched, or pitted? Rounded edges and faint trade marks and serial numbers may indicate a refinished gun, while a very worn or rusty gun may have to be refinished by you, the purchaser.

Damaged screw heads and other signs of tinkering warrant particularly close examination of a potential purchase.

Check the muzzle for nicks or burrs that can hurt accuracy. While a gunsmith can recrown a

This article first appeared in American Rifleman

barrel to restore accuracy, this repair would justify a lower price.

Examine the bore, preferably with a bore light or mirror. If the lands are rounded, the gun has likely seen considerable firing. Any minuscule lumps in the grooves may be lumps of lead or metallic fouling, indicating poor cleaning practices by the previous owner.

Any dark, irregular patches that resist a pre-purchase cleaning are likely signs of rust or pit-

Prospective buyer should know critical areas to inspect in used revolver, and has every right to insist on live-fire testing. Tips in this article will help you distinguish peach of a gun from a lemon.

SOME BASIC REVOLVER CHECKPOINTS

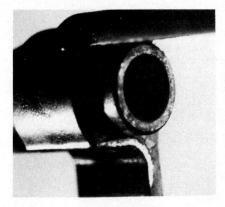 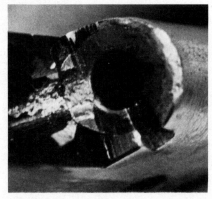 

Barrel throats on revolvers are subject to belling, splits and gas cutting. Take close look at this area (left) and inside the cylinder. Battered barrel face (center) and rough cartridge ramp should discourage purchase. Pocket magnifier allows really close inspection. Revolver's ratchet or "star" (right) should be examined for wear. Also check condition and function of bolt notches in cylinder.

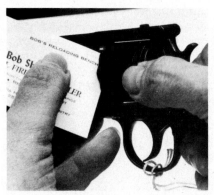 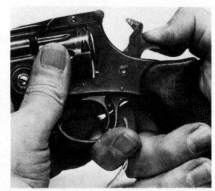 

Business card can give rough check of barrel/cylinder gap. If typical .010-inch-thick card can slide inside, gap is excessive. Revolvers should be cycled, checking timing and locking sequence. With hammer cocked, make certain it won't push forward. Crane misalignment usually can be repaired, but it adds to revolver's cost and is hint that firearm has been abused.

ting, another mark of improper care. A bad bore can be rectified by replacing the barrel, but that's rarely very economical.

The barrel should be checked to assure that it is firmly attached to a revolver's frame, and that its throat is free of cracks or deformities. Whether pinned or screwed into place, the barrel should be rigid, even under the maximum torque you can apply with bare hands. Looseness often indicates an improperly installed replacement barrel. And loose or not, a barrel with a cracked or belled throat should always be avoided.

Stocks or grips are the easiest, and cheapest, items to replace, and should have relatively little importance in setting a price, though ugly grips often drive away the inexperienced shopper.

Of far greater importance are the sights. If they are fixed, do they show any signs of being altered, or have they been bent or burred? Nicks and burrs can easily be dressed and polished out, but if the sights are bent, or if metal has been filed away to target the gun for a specific load, repair can be a costly problem.

If the rear sight is adjustable, the sight blade should be checked for looseness and the adjustment screws for proper movement. A simple test is to apply pressure in one direction with the thumb, then use the adjustment screws to move the blade against it. If the movement is sluggish or loose, or fails to occur at all, then another possibly costly repair is definitely in order.

On revolvers, the cylinder should be removed or swung open and the chambers closely checked. They should be clean and smooth throughout, since any roughness, burrs, or pitting may make for difficult case extraction, while at the front of the cylinder, any roughness or erosion in and around the chamber mouths may mean the gun has seen hard use, possibly even with corrosive ammunition in the case of older guns.

The ratchet or "star" at the cylinder's rear should be reasonably sharp and well-defined; so, too, should the cylinder slots or bolt notches located around its periphery. If either is badly worn, the chambers may not align correctly with the bore, and the gun may have to be retimed, a costly and tedious procedure many gunsmiths refuse to undertake, except at an hourly rate.

It's important to use proper terminology when discussing repairs with a gunsmith or ordering parts. What Colt calls a bolt, for example, Smith & Wesson terms a cylinder stop and Ruger a cylinder latch. The shopper who is unfamiliar with revolver terminology should consult the *American Rifleman*, June, 1985, p. 62, or books on firearms assembly like NRA's *Firearms Assembly 4*, an especially valuable reference that covers both autoloaders and revolvers.

With the cylinder replaced, a specific check of of the timing can be made by cocking the gun over each of its several chambers. The timing should be such that, as each of the chambers rotates into position, the cylinder stop should pop up into the slot an instant before the hammer engages the sear. This can usually be heard as two distinctly audible clicks occurring in quick succession, one for the cylinder stop, the other for the hammer nearing full cock.

If only one is heard, either the cylinder stop or bolt has failed to engage the cylinder, or slot engagement has occurred just as the hammer has reached full cock, allowing the final click to mask the sound of the cylinder stop.

In the course of working the action, each time the hammer is fully cocked, the cylinder should be checked for rotational play. Since there is no way to quantify this movement, it can only be said that it should not be a lot. Most important, moderate hand effort in either direction must not rotate the cylinder out of lock.

Cylinder and bore alignment can be checked with the help of a range rod, available from Brownell's (Rt. 2, Box 1, Montezuma, IA 50171). The range rod is inserted from the muzzle, and will strike the front of the cylinder if it is grossly misaligned. It should be noted that a range rod doesn't guarantee perfect alignment; it merely detects large-scale misalignment.

While different revolver brands have different standards for rotational play, in no case should movement be much.

Checking the action means checking the trigger pull, too. This is an area where home hobbyists often foul up revolvers, then try to stick the unwary purchaser with damaged goods. If the pull is unusually light or heavy, beware a budding gunsmith. When at full cock, finger off the trigger, the hammer should not drop when firm thumb pressure is applied against it.

The revolver's action should function smoothly, with no feeling of roughness or hesitation in its movement. Both the hammer and trigger should be free of excessive side play, and also should show no signs of excessive wear on their sides that might indicate improper assembly. The trigger should move smartly into the ready position when released after the hammer is dropped.

At the same time, the hammer should return to its rebound position. (Certain older designs don't feature rebounding hammers, and those should be considered mainly as collector's items).

Whether attached to the hammer or fitted within the frame, the firing pin's tip should be smooth and round, and its fit through the frame should be snug but unimpeded. A battered or corroded pin, or one with improper clearance, can be the source of problems ranging from outright failure to fire to occasional misfires to pierced or extruded primers.

The frame should be checked for any signs of ridging or burring around the firing pin hole, a condition that often attends a faulty firing pin, and one that signifies a gun that has been fired or dry-fired excessively.

A crucial item to check on a swing-out cylinder gun is the fit of the crane or yoke into the frame. The crane should fit into its recess in the frame with no visible gap between the two. Anything more than a hairline here will throw the cylinder out of alignment with the bore.

The problem may be caused by nothing more than dirt in and around the frame, especially at the pivot point. All too often, though, the revolver's crane has been bent by a previous owner who, having seen too many old movies, made a practice of flipping the cylinder open and closed one-handed.

This can also bend the ejector rod; check for a bent rod by spinning the cylinder while observing the rod for eccentric rotation. A bent ejector rod may show up in the amount of effort needed to close the cylinder or to disengage the latch that releases it. While there may be other causes for a bent rod, it is often attended by a bent crane, and the shopper would be wise to look for another gun.

SOME BASIC AUTO CHECKPOINTS

Test semi-autos to make sure they won't fire out of battery. With slide slightly retracted (left) cocked gun should not fire. Check grip safeties by cycling empty gun and squeezing trigger without depressing safety (center). Hammer should not drop. Damaged firing pins (right) are hint of abuse. Examine both ends, where possible, looking for deformation that may require repair.

  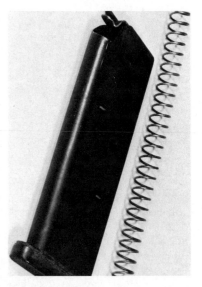

Faulty magazines or recoil springs may be to blame when smokestack jams occur (left). This is among most irksome semi-auto woes. Disassembling autoloaders permits look at critical sliding surfaces and good examination of barrel bore and chamber. Magazines and recoil springs are most common causes of autoloader failure. Test firing ensures that they function right.

The ejector rod shold move freely, and the ratchet or ejector should seat fully in the cylinder.

After checking the fit of the crane in its recess, the gap between the barrel face and cylinder should also be checked and measured. Measure the gap between barrel face and cylinder with the revolver in battery; that is, with the hammer fully down in firing position.

The gap between cylinder and barrel face should be around .005 inch, ideally. Even many new revolvers fail to achieve this, however. As a practical matter, anything up to about .008 inch is acceptable. One simple test is to try inserting a common business card, about .010 inch thick. If the card passes, the gap is too large.

This is also a good time to check the cylinder for end play. It should not be an appreciable amount.

After these checks, which take less time to perform than to read about, an acceptable gun should still be test-fired before the sale is completed or a conditional sale negotiated. Still, these checks should lessen the chance of buying a "lemon" you'll regret owning.

Autoloaders tend to have fewer closely fitted parts than do revolvers, but on the other hand, they require live firing for a satisfactory and reliable function test.

One major difference between revolvers and autoloaders is that, while very few wheelguns have safety devices, semi-automatics may have several. The 1911 Colt, for example, has both a thumb safety and a grip safety (Series 80 models add a firing pin safety), while many other autoloaders have a magazine safety. Don't even look at a self-loading pistol unless you clearly understand how it and all its safeties are supposed to work. Otherwise, examine it only with an instruction manual in hand.

Another vital safety feature of auto-loaders is the disconnector (sometimes interrupter), a part designed to prevent a multiple discharge with one pull of the trigger. With the gun cocked and empty, the slide or breechblock should be drawn to the rear about ⅛ inch and the trigger firmly depressed. If the hammer or striker falls, the disconnector may be defective and require repair or replacement.

For an obvious reason, an autoloader must remain cocked as the slide or breechblock moves forward into battery. This feature should be checked by drawing the slide of the unloaded gun to the rear and, with the finger *off* the trigger, allowing it to travel forward under the force of the recoil spring. Then, with the trigger depressed, the test should be repeated. If, in either case, the hammer or striker falls, the sear engagement may be faulty and in need of repair.

Of equal importance is the thumb or manual safety. If, when engaged, it fails to keep the trigger from releasing the hammer or striker, the safety is defective. If the gun is also equipped with a grip safety, familiar to anyone acquainted with the M1911, it should be checked by cocking the gun and pulling the trigger without depressing the safety. If the hammer or striker falls, the grip safety may need repair or may have been deactivated by an earlier owner.

If the pistol is a double-action automatic, engaging the thumb safety may allow the hammer to fall, but only to a ready position. This can be checked (in guns of .30 caliber or larger) with the familiar pencil test. Drop a pencil eraser-first down the bore and release the safety. If the safety doesn't work, the firing pin will strike the pencil, popping it up the barrel. You may have to hold the pencil against the breech face in .45 caliber pistols. This test is worth performing especially on guns that may have been the victims of home gunsmithing.

Some guns are equipped with magazine safeties intended to prevent firing without a magazine in place. These safeties are often removed for service use and practical pistol competition. If the magazine safety has not been removed, it can be checked by removing the magazine, cocking the gun, and pulling the trigger.

Some modern autoloaders are equipped with passive firing pin safeties that help prevent accidental discharge in case the gun is dropped. These lock the firing pin except when the trigger is depressed. Check this feature by cocking the hammer, then pressing forward on the head of the pin with a pencil point or small punch. If the pin moves forward freely, the safety needs, repair or replacement.

No matter the make of the gun, the firing pin itself should be examined closely. Its tip should be smoothly rounded, while its head should be free of any burring or peening that might interfere with movement. The firing pin return spring should also be checked for kinks or breaks.

The sliding parts should be checked for damage or excessive wear. Cycle the slide or breechblock several times, checking for binding or excessive looseness. Then field-strip the gun and check for nicks, burrs, gouges, or rust spots that might impede reliable function. While autoloaders need a certain amount of clearance between sliding parts, looseness of fit to the point of being sloppy will detract from accuracy.

The magazine is absolutely vital to the proper operation of a self-loading pistol, and the body and feed lips should be checked for obvious flaws. Of course nothing substitutes for range testing with live ammo here. Every experienced pistolman has had horrid-looking magazines that fed flawlessly, and others that looked brand new and did nothing but jam.

Replacement magazines are available for just about any gun ever made, but one for an obscure sidearm can be costly, and the gun's asking price should be adjusted accordingly.

Again, a live-fire test is always the best way to prevent getting stuck with someone else's problem. A seller who adamantly refuses to allow a range test must be suspected of having something to hide. In buying a used pistol, as with any other product, the watchword must remain "let the buyer beware."

Ruger's .22 Pistol . . .
The Evolution of a Classic

Bob Milek

Ruger's Standard Automatic Pistol, introduced in 1949, was the foundation on which America's great gunmaking firm, Sturm, Ruger and Company, was built. Bill Ruger is a firearms design genius and an astute businessman. However, I doubt if even Bill visualized just how important the little automatic pistol would become both to his business and to American shooters.

For 33 years the Standard Automatic Pistol (automatic is really a misnomer because the pistol is a semi-automatic, not full automatic), Bill Ruger's first gun, remained a staple in the company's product line. In 1951 a target version, the Mark I Automatic Pistol, was introduced, and this was followed later by a bull-barrel version. The Mark I was adopted as the training arm of the Army and Air Force, a move that didn't hurt its popularity at all. Still, Ruger's semi-automatic rimfire pistol was never conspicuous by its presence. It was just there—selling and making its owners happy. And it hung in there while many of its competitors, guns with a heady reputation in competition circles as well as with hunters, fell by the wayside.

Then, in January of 1982, Ruger announced the end of all models of the Standard Automatic Pistol. In its place would be a new .22 rimfire pistol—the Ruger Mark II. The Mark II has been with us for five years now, and to attest to its success it's safe to say that Ruger owns the .22 semi-automatic pistol market in America. The Mark II has competitors, but the bulk of the market belongs to Ruger.

Today three models of the Ruger Mark II are available: the Standard Model which has fixed sights and your choice of 4¾- or 6-inch barrel; the Mark II Target with adjustable sights and a 6⅞-inch barrel; and the Mark II Bull Barrel which sports adjustable sights and your choice of a 5½-inch or a 10-inch heavy bull barrel. All are available in either blued steel or stainless.

Just what is it about the Mark II that makes it so popular, and how does it differ from the original Standard Automatic Pistol? In answer to the latter question, you might be surprised. The differences between the new and old models are subtle indeed, but they represent changes that were necessary to make the gun one of the world's outstanding .22 semi-automatics.

The Mark II pistol is of blowback action design. In other words, the barrel and breechblock are not locked together for firing. Instead, inertia and pressure from the recoil spring hold the breech bolt against the face of the barrel until gas pressure from the firing of the cartridge builds sufficiently to force it rearward. This pressure is reached after the bullet leaves the barrel, thus there is no danger of the breech bolt opening prematurely and spewing hot gas all over the shooter.

Now is probably a good time to delve into exactly how the Ruger Mark II works. With the breech bolt held open by the bolt stop, insert a full magazine into the bottom of the grip frame and push it in until the magazine latch engages. Pull back slightly on the breech bolt and release it. As it moves forward, powered by the recoil spring, it strips a round from the magazine and feeds it into the chamber. The pistol is now ready to fire. Aim at your target and pull the trigger. As the trigger moves rearward, it drags the trigger link, called the disconnector by Ruger, forward. In turn, the trigger link pulls the sear out of engagement with the notch on the internal hammer. As soon as the sear clears the notch, the hammer is powered forward by the mainspring and strikes the intertia-type firing pin, driv-

This article first appeared in Guns & Ammo

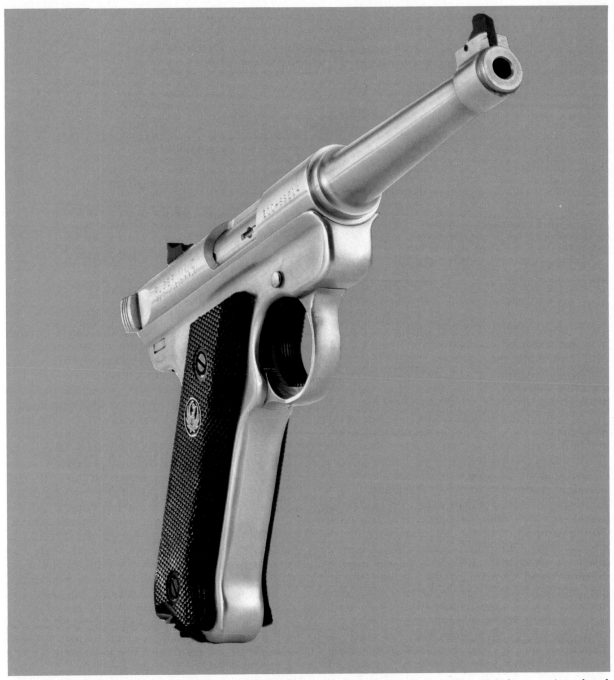

Ruger Mark II, available in three versions, is direct descendent of company's first .22 pistol that was introduced nearly four decades ago. Though differences in newest model are subtle, author believes they make it one of world's outstanding .22 autos.

ing it forward to strike the rim of the chambered round, causing it to fire. The firing pin is immediately drawn back into the bolt by the spring.

Upon firing, gas pressure from the burning powder exerts pressure both forward and rearward. Forward pressure drives the bullet out of the case and down and out the barrel. Rearward

pressure overcomes the weight of the breech bolt and drives it rearward. In this movement of the bolt, the fired case is extracted and ejected, the recoil spring is compressed, and the internal hammer is cocked.

The bolt continues to move rearward until it contacts the bolt stop. Then the recoil spring de-

Various types of ammo, even hyper-velocity .22s, fed and functioned reliably in Mark II. No malfunctions.

compresses, driving the bolt forward again. As it moves it strips a fresh round from the magazine and feeds it into the chamber. The bolt's forward movement stops when it makes contact with the face of the barrel. Pulling the trigger again repeats this cycle until the last shot in the magazine is fired. After the last shot, the bolt stop jumps up, forced to such position by the magazine, and holds the bolt open so that you are aware that the last shot has been fired.

The operation of the Mark II pistol is just as simple as it sounds. However, realize that there's a delicate balance there between the weight of the breech bolt and the pressure of the recoil spring as opposed to the pressure generated by the .22 Long Rifle round being fired. If ammunition is used that generates pressures below those deemed standard by the industry, the action may not cycle. Likewise, unreasonably high pressure could cause the action to open prematurely or cycle so quickly as to cause a malfunction. The latter case is something I've never experienced, and I doubt any shooter ever will. However, I have on occasion experienced malfunctions due to low chamber pressure. These can normally be avoided by simply using quality .22 Long Rifle ammunition. Bargain-basement stuff may give you trouble, as will any attempt to use .22 Short or .22 Long ammunition. The Mark II is designed to function only with .22 Long Rifle ammunition.

The bolt stop on the Mark II pistol is one of the subtle changes made when the old pistol was redesigned. On the old guns there was no way to tell when the last shot was fired, thus the shooter was forever squeezing off on an empty chamber. Not only was this annoying, but after

the firing pin thunked into the barrel enough times—striking where the rim of the case should have been—a burr was sometimes raised on the face of the barrel which could prevent complete closing of the bolt. This is all history with the Mark II. It has a bolt stop that functions beautifully. When the last shot has been fired, the magazine forces the bolt stop up so that it catches and holds the bolt in the rearmost position. The bolt can be closed after removing the empty magazine by either pushing down on the outer bolt stop thumb piece located on the left side of the frame at the front top of the grip panel or simply pulling rearward on the bolt, then releasing it. The bolt stop is spring-loaded downward, so once you pull back on the bolt and release its pressure against the stop, the stop moves down out of the way. The latter method of bolt closure is recommended when a loaded magazine has been inserted into the gun.

The magazine and magazine latch in the frame are another important area of change between the old Ruger pistol and the new Mark II. The magazine latch on the new gun is wider, affording a more secure grip on the magazine, and the magazine itself has been redesigned so that it loads easier, is less prone to malfunctions, *and* is more easily removed from the gun. The upper lips of the magazine, the follower, and the follower spring all have undergone considerable redesigning. The magazines in all of my Mark II pistols "jump" down when the latch is released, making them much easier to remove. On the old guns a man could dig around forever before he managed to get a grip on the bottom of the magazine so he could pull it out.

A very important addition to the Mark II is a

modified safety that allows the pistol to be unloaded while the safety is on. If there's one thing about most semi-automatic pistols that scares me, it's the fact that once you pull that trigger, the pistol is again reloaded, and to unload it you must retract the breechblock with the gun in full fire position. The safety on the Mark II is positive. When the safety button, located on the upper rear left side of the frame, is pushed up into the "Safe" position, the inner portion of the safety moves down and locks the sear in place. You can pull on the trigger all you want, but you can't jerk the sear out of engagement with the hammer as long as the safety is on. Now, to make things even better, the breech bolt can be operated with the safety on, thus the loaded round in the chamber can be removed.

The only other major difference between the old Ruger semi-automatic .22 and the new Mark II is one that few shooters will ever be concerned with. It has to do with the method by which the trigger pivot is retained. In the old pistol a lock washer retained the pivot and somewhat complicated disassembly of the firearm. A piano-wire spring is used as the retainer on the Mark II, greatly simplifying the disassembly job.

There are, of course, some other minor changes between the old and new Ruger guns. There is a more precise fitting of such things as the trigger and magazine; a more attractive trigger, .365 inch wide, is used on the new gun; and shallow scallops have been cut into the sides of the receiver at the rear to allow the shooter to get

a better grip on the cocking lugs of the bolt. Nice changes, but nothing to get excited about.

If there's anything about the Mark II pistol that needs improvement, it's the trigger pull itself. On my three test guns the trigger pulls are 3 pounds, 4½ pounds, and 5 pounds—only the 3-pound pull being satisfactory to me. However, remember two things about the Ruger Mark II. First, it's a semi-automatic. This means that there's a lot slamming and banging going on, particulary during the rapid-fire sequence. If the trigger notch on the hammer and the sear itself were honed to produce a superb pull weight of say 1½ or 2 pounds, we'd run the risk of having the pistol fire when it received only a slight jar. Therefore, the sear notch on the hammer, as well as the engagement surface of the sear, must be healthy enough to prevent any dangerous malfunctions. Second, the Mark II uses a trigger link—a long bar of steel connected in front to the trigger itself, in the rear to the sear—to drag the sear out of contact with the hammer. Any time you involve a link of this type with the trigger system, you're going to get some "softness" that isn't associated with a more basic system such as we find on revolvers. Anyway, the Mark II trigger can be improved, but the work should be undertaken *only* by an expert pistolsmith. For the average shooter to attempt to hone the hammer and sear of the Mark II in his basement workshop is to invite disaster.

The adjustable sights on the Mark II Target and Bull Barrel models are excellent. Up front there's

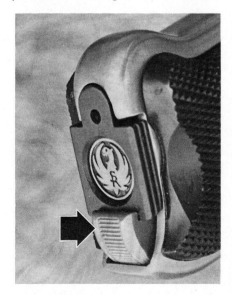

Wide, hefty latch (arrow) makes magazine removal faster and easier with new Ruger .22 pistols. At near right is Mark II with 4¾-inch barrel and fixed sights; next to it is target grade with adjustable sights and 5½-inch bull barrel. See text for other barrel lengths now available.

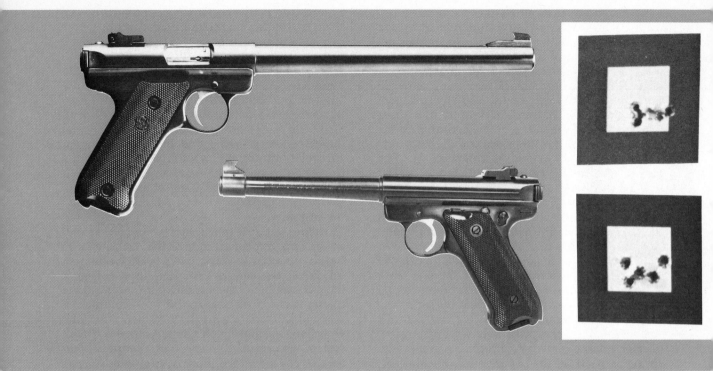

Ruger makes several target models with adjustable sights. Pictured here are 10-inch bull-barrel version and target pistol with 6⅞-inch barrel. Both targets pictured to right of guns were fired with 5½-inch bull-barrel pistol, hand-held over sandbags at 25 yards. Upper group was produced with Federal high-velocity hollow-points, and group below was produced with Winchester high-velocity hollow-points.

a ⅛-inch-wide blade. The rear unit, set in a dovetail milled in the top of the receiver, is fully adjustable for windage and elevation. Having recently run up against a problem that comes to all of us with age—the need for reading glasses—I really appreciate the large, square notch Ruger cuts in the rear blade. It's .080 inch deep and .090 inch wide, affording a lot of daylight on both sides of the front blade. I need a lot of daylight to properly center that front blade in the rear notch.

One of the really attractive features of all Mark II pistols is the manner in which they can be fieldstripped for easy cleaning, negating the need to completely disassemble the pistol. After removing the magazine and checking to be certain the pistol is unloaded, close the bolt and pull the trigger to release tension on the mainspring. Insert a blunt instrument such as a screwdriver in the oval under the mainspring housing latch, located on the rear of the grip frame. Pull the latch down, then swing the entire housing outward and pull down to remove it from the frame. Point the muzzle up, and the bolt will drop out the rear of the receiver. Using a plastic or rubber hammer, strike the rear of the receiver a sharp blow toward the muzzle. This drives the barrel/

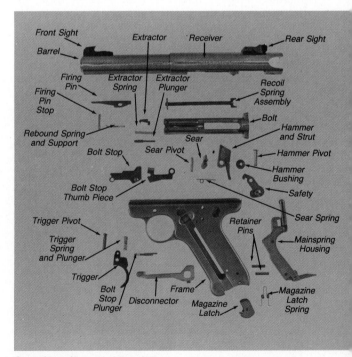

Complete disassembly of Ruger target pistol, such as this, is seldom necessary and is not recommended for inexperienced shooters.

RUGER .22 PISTOLS—ACCURACY AND VELOCITY

Ammunition	4¾-INCH BARREL		5½-INCH BARREL		10-INCH BARREL	
	Vel. (fps)	Avg. Group Size (Inches)	Vel. (fps)	Avg. Group Size (Inches)	Vel. (fps)	Avg. Group Size (Inches)
CCI Green Tag	906	1.70	926	1.40	1,008	1.00
Fed. Pistol Match	945	.61	946	.57	1,068	.62
Eley Pistol Match	915	.68	944	.50	1,023	.55
RWS Pistol Match	899	1.04	910	.80	975	.90
CCI Mini Mag HP	1,059	1.50	1,097	1.30	1,222	1.02
Fed. HV HP	1,082	.80	1,128	1.10	1,225	.61
Lapua HV HP	1,030	1.00	1,100	1.10	1,203	1.40
Rem. HV HP	1,041	1.10	1,072	.90	1,149	.79
Win. HV HP	1,057	1.30	1,080	1.00	1,203	.78
CCI Stinger	1,191	1.05	1,266	1.20	1,395	.87
Fed. Spitfire	1,185	1.80	1,212	1.50	1,350	1.65
Rem. Yellow Jacket	1,162	1.40	1,187	1.60	1,297	1.48
Win. Super-Max	1,123	1.20	1,100	1.25	1,330	1.10

receiver forward so that it detaches from the frame. The bore can now be cleaned from the rear and the receiver is easily reached for cleaning. To remove grit and old lube from the trigger, sear, and other parts located in the frame, remove the grips, place the frame in a pan of solvent and slosh the frame around vigorously. Remove the frame, allow it to stand until all of the solvent has evaporated, then relube the moving parts and pivots.

I tested three Mark II pistols for this article—a Standard Model with a 4¾-inch barrel, a 5½-inch stainless steel Bull Barrel model, and a blued 10-inch Bull Barrel model. Let me say right off that my use for these pistols is 99 percent hunting. As I've often said before, I have no use whatsoever for fixed sights on a hunting pistol. Therefore, I topped my Standard Model with a Thompson/Center 3X pistol scope set in the B-Square mount made especially for this pistol. Both bull-barrel guns were tested with the factory open sights. I tested a variety of hunting ammunition, both high velocity and hyper velocity, as well as a few brands of pistol match ammunition for both velocity and accuracy.

Do the Mark II pistols shoot? You'd better believe it! Five-shot groups at 25 yards averaged ¾ to 1 inch with several brands of hunting ammo, and with Eley Pistol Match my 5½-inch Bull Barrel Mark II consistently printed five shots in ½ inch at 25 yards. I'll guarantee you I can't see to shoot any better than that, and I don't know what more a man could ask from a pistol. Of the hyper-velocity ammunition I tried, CCI Stingers turned in not only the highest velocity, but the

best accuracy. Federal Spitfires, Remington Yellow Jackets, and Winchester Super Max were all comparable in the accuracy category. For small-game shooting where precise bullet placement is a prerequisite for quick kills and minimum meat damange, I'll stay with high-velocity hollow-point ammunition in my Mark II pistols, but for hunting small varmints, I like the hyper-velocity ammo.

The accompanying table gives all of the velocity and accuracy data I recorded with various .22 Long Rifle ammo in my Mark II pistols. The accuracy with all three of the pistols I tested was excellent, hinting that barrel length doesn't appreciably affect accuracy. However, you'll notice that it does have a noticable effect on velocity.

Malfunctions were nonexistent with all three of my test pistols. Ammo fed from the magazine into the chamber without a hitch, extraction was perfect, and all other functions of the pistols were 100 percent. I was especially pleased that the Mark IIs handled hyper-velocity ammunition so well. Often a semi-automatic pistol won't feed truncated-cone bullets properly. These are no problem for the Mark II.

Seldom does a man get to work with guns that perform so well throughout every single category of his tests. I've used my Mark IIs in extremes of heat and cold and under both dusty and wet conditions. They take it all in stride and go right on shooting. This, plus superb accuracy, are the reasons I consider Ruger's Mark II pistols to be one of the best buys for the money in the firearms marketplace today.

PART FOUR

HISTORY AND BLACK POWDER

Adolph Toepperwein, Exhibition Shooter

Col. Charles Askins

DURING the early days of this century there was a tight little coterie of shooters who delighted their audiences with feats of gunning skill no longer seen. The stalwarts fired mostly shotguns or rifles with shot loads, and their target was usually the glass ball. There was a notable exception, and this was a Texan named Adolph Toepperwein.

Most of his feats were accomplished with the .22 rifle with the conventional lead bullet. He first fired the Winchester Model 1903 autoloader, and later on, when the successor to this sturdy number came along, the Model 63, he swung over to it. Both were .22 caliber, although the '03 used a special .22 Automatic round, not the .22 Long Rifle.

Toepperwein was of German extraction. His father, Fred Toepperwein, was a farmer and a gunsmith, and one of the son's proudest possessions was a rifle designed by his pater. Patent rights were sold to Winchester, although nothing was ever done with the pilot model. The family home was on a small ranch near the present town of Leon Springs, about 25 miles north of San Antonio.

Tall and angular with steely blue eyes and a direct way of talking to the other fellow, Toepperwein had all the makings of an early-day Texas Ranger. Born in 1869, when the Comanches and Apaches were still a hazard to those settlers who lived outside the pueblos, the young Toepperwein often saw roving bands of the horse-mounted redskins. If there were any fights with the raiders, he never spoke to me about these encounters.

The young Leon Springs citizen was an inveterate hunter. Settlers lived on wild game in those days, and Adolph kept the table in bountiful supply. His first rifle was a .40 caliber muzzleloader, but later he swung over to the Winchester '73 in .44 WCF with 24-inch octagon barrel.

At the age of 18 the bright lights beckoned, and young Toepperwein descended on San Antonio. The population in the 1890s was around

This article first appeared in American Rifleman

A · D · O · L · P · H
TOEPPERWEIN

EXHIBITION

SHOOTER

Toepperwein, then 92, posed with Askins in 1962.

10,000. Ad, as he was called, found a job as a sales clerk in a clothing store on Commerce Street. Besides his extraordinary ability with the rifle, exhibited at an early age, he was especially talented as a cartoonist. He soon gave up his career in the mercantile business and moved over to the San Antonio *Express*, the only daily newspaper, as a cartoonist. In those distant days—it was in the mid-90s—drawings were first made on a chalk plate. These ancient sketches are still preserved in the *Express* archives.

For diversion our Texan continued to shoot at every opportunity. He had the Winchester Model 1890 slide-action .22, and his bosom buddy, Louis Heuermann, tossed the pebbles, cans, blocks of wood, and whatever else offered a suitable aerial target. Toepperwein was not much interested in stationary targets, and I never knew of him firing on a bull's-eye. The mark had to be moving if it was to hold his interest.

The Sells-Floto Circus came to town one day and there was a trick shooter who did his stuff. Emulating Buffalo Bill Cody, he galloped around the ring bursting balloons as he rode. His rifle was the Model '73, but it had been bored out and fired only .44 shotloads. "I can do that with the .22 Long Rifle and lead bullets," Adolph exclaimed. Somehow word got to the circus owners, and they asked for a demonstration. The record does not show whether Toepperwein ac-

tually galloped around the ring and plunked the globes with his trusty pump-action .22 but this demonstration was sufficiently impressive to give him the job. The circus went to Mexico, and Adolph went along.

A short time later—in 1894 to be exact—a long, hungry, 6-foot Texan came drifting up out of Mexico. He had been down there with the circus for almost a year putting on rifle and pistol shooting exhibitions. Without fanfare or publicity he broke 955 out of 1,000 clay disks, each 2¼ inches in diameter, with a .22 rifle. However, since most trick shooters were using glass balls, Toepperwein's accomplishment didn't constitute a record.

A short time after his preliminary attempt on the clay disks, Top tried again, and this time he pulverized 987, followed in a short time by 989. Then, by way of a little diversion, he cracked down on 1,500 regulation claybirds and hit all of them. Again, this was with a .22 rifle, the first 1,000 targets thrown at 30 feet rise, the last 500 at 40 feet.

By the turn of the century Toepperwein was in the hire of Winchester. He was to shoot exhibitions for the New Haven firm for 55 years. In 1903 he married Elizabeth Servaty, and in no time at all he made her a worthy partner in the exhibition business. He nicknamed his little frau "Plinky," and so she was known for all her professional career.

Mrs. "Top" became an extraordinarily talented trapshooter and ran many 100s straight. She first shot the Winchester Model 97, and then, when the Model 12 came along, she swung over to it and shot this pump repeater throughout the remainder of her long career. In 1916, firing at the Montgomery, Alabama, Country Club, Mrs. Top shot at 2,000 regulation clay targets during a period of 3 hours and 15 minutes and broke 1,952 clays. She had long runs of 280, 139, 111, and 106 targets without a miss.

Toepperwein traveled through every state and all the territories, usually by train, and carried along his several firearms, his considerable quantities of ammunition, and, just as importantly, his multitudinous targets. He was met usually at the railroad station, by the local sporting goods dealer who, of course, sold Winchester shooting irons.

This worthy set up the affable Texan in the front of his store to visit with the local patrons and answer their silly questions and listen to their impossible shooting tales. That afternoon, he would move outside the town or city and put on his shooting extravaganza.

This, remember, was more than a half-century ago, and unlike today there was plenty of open country beyond the city edge. Not only that, but

"Top" was photographed as he sat on enormous heap of wooden blocks shattered in his 1907 record run. With his Winchester .22, he fired at 72,500 of these aerial targets—and missed only nine!

there was utterly no concern as to where the bullet would land.

Ad Toepperwein shot everything well. He would turn a back flip, the Model 12 scattergun in hand, and break five clay targets which he had tossed skyward before he did the back somersault. He laid the Model '03 rifle on its side, pulled the trigger, and when the empty was ejected, he spun the muzzle around and neatly hit the spent casing. He aimed a .22 revolver over his shoulder, using a mirror, and aimed a second six-shooter to the front, pulled both triggers simultaneously, and struck a tin can behind and another in front.

Though Top didn't smoke, he would produce a cigarette, light it, and stick it in his mouth, and Plinky would neatly shoot off the ash with a .38 revolver. He would throw two cans in the air and hit both of them, then three cans and finally five cans, all punctured before they struck the ground.

A real crowd pleaser was the Uncle Sam or Indian chief in full feathered headdress that Top drew with .22 bullets. This exhibition was done on a piece of sheet tin, and the marksman fired from about 20 feet. The head was always slightly larger than normal. There were no marks on the tin, no lines to be followed with the bullets, no spot from which to begin.

Toepperwein simply commenced shooting, and at a rate of about one shot per second he put 350 bullets in the artistic effort in about six min-

utes. These "drawings" are notable for the fact that there is not a single shot out of place or out of line. I have a Toepperwein Indian as a much prized souvenir of the remarkable skill of this amazing marksman.

Ad Toepperwein had some competitors who sometimes shot the shotgun but also used the .22 rifle. Chief among these was Doc W. F. Carver, and another was Adam Bogardus. Along with them was the renowed Annie Oakley. Of somewhat lesser frame was Dr. Ruth and a man named B. A. Bartlett.

The usual target was a glass ball some 2½ inches in diameter. This bauble was tossed aloft by a singularly well trained individual who contrived to heave the spinning sphere into precisely that position where it could be most easily hit. The distance from gun to thrower was 25 feet, and the mark flew skyward about 20 to 25 feet. The skilled marksman essayed to hit it just at the apex of its flight.

Doc Carver, who explained to one and all that he was the world champion trick and fancy shooter, got all wound up one day and commenced firing at the glass balls. His rifle was the Winchester Model 1890 chambered for the .22 Short. There were 10 rifles in all, and Carver kept a squad busy cleaning them. The old Lesmok cartridge was a dirty number, and the outside-lubricated bullet presented still another problem.

After 60,016 of the glass targets had been fired on, the score showed Doc had missed 4,865. He had shot from 11 a.m. each day until 11 p.m. every eve, for a full 10 days. It was a grind.

Later on the inimitable Carver again essayed the stint, and on this second try he missed only 650 targets. Based on this demonstration, the modest shooter—he was not really a medico at all, despite the title of Dr.—laid a more insistent claim to the world championship.

This was lovely until a relative unknown, B. A. Bartlett of Buffalo, New York, oiled up his Winchesters one day and commenced shooting. Some 144 hours later, firing at the regulation 2½-inch glass ball, the man from Buffalo had shot at 60,000 spheres and missed only 280. The record stood for a number of years.

During mid-December, 1907, Toepperwein moved onto the San Antonio Fair Grounds with 10 Winchester rifles, the Model 1903 autoloader, and 50,000 cartridges. Also on hand were seven wagonloads of sawed wooden blocks, each cut precisely to 2¼ x 2¼ inches. Now just why our Tejano elected to fire at 50,000 targets and not the usual 60,000 we cannot say. Too, it is something of a mystery why he did not use 2½-inch glass balls which had been Carver's choice.

After 10 days, during which time the Winchester marksman had shot 12 hours daily, he had fired on all seven wagonloads of blocks, and some 22,500 had been shot at twice. Top had six boys throwing for him, and these lads were exhausted. He had two judges and a scorekeeper, and they also admitted they were well tuckered out. After the stint the score showed that some 72,500 blocks had been fired on and only nine had been missed. There was one long run of 14,500 without a single miss. When Toepperwein ran through his 50,000 rounds, which had been the original goal, there was a great deal of mad scurrying to find additional cartridges.

The supply of blocks became exhausted, and Toepperwein commenced on the larger pieces. As these fragments were burst asunder, he salvaged what was left of the splinters and shot at them.

Toepperwein, in telling me about the shooting many years later, said he had to be put to bed after it was all over. "I would get my arms up as high as my shoulders, and then I could not get them down again."

The record was to stand for more than a half century.

Exemplary though it might have been, the Toepperwein record was pretty well forgotten. Exhibition shooters came and went, but none essayed the 72,500 targets. That is, none save a fellow named Tom Frye. This hombre was not a trick and fancy shooter at all but a salesman. He sold Remington guns and cartridges and shot a great deal of regulation trap as a professional.

Sometimes he did exhibition shooting, but it was not a regular thing. He got little encouragement from the UMC because the company had looked into the cost of maintaining a shooting shark on the road and the cost was simply too high. Frye could have all the rifles and quantities of ammo he needed, but that was the limit of it.

Despite the lack of encouragement from the Bridgeport office, our boy had a secret plan in mind. When the .22 caliber Remington Model 66 nylon autoloader bowed onto the stage in 1959, Frye acquired two of the new rifles and, taking his annual vacation leave, moved out to Reno, Nevada, where he got together with Newt Crumley, owner of the Holiday Hotel and game farm.

As indication of the enthusiasm the hotel man had for his friend Frye, not to mention his faith in the Frye shooting ability, Newt not only had the wooden blocks sawed to the regulation 2¼ inches, but also purchased the .22 cartridges for the forthcoming shooting grind.

The time was mid-October, 1959, and the plan was to fire not on the 72,500 targets which Ad Toepperwein had busted but on a full 100,000 blocks. Frye was to shoot at 10,000 blocks every day for 10 days. He got under way each morning at 8 a.m. and continued to shoot until it was too dark to see.

Despite his good intentions, he banged out only 3,000 targets the first day. Because of the crowds, the newspapermen, and the TV cameras, he fell a bit short of the planned 10,000 squares.

While the rifles were the Remington nylon gun, Frye fired Peters cartridges. "I thought that was only fair," he explained to me, "since I sold the shells." While he got a slow start that first day, he quickly made up for it. By dusk of the third day he had 27,310 blocks. By the sixth day the tally showed 43,725 targets fired on and only two escapees. When 63,500 blocks had been tried, there were three misses.

When Frye fired at the 72,501st target—and hit it—he had missed only three times. It had taken him 12 days. That day he shot at 9,500 blocks and sometimes worked up to speeds of 40 blocks per minute.

While it was the original plan to quit after 10 days, Frey continued to fire through the 13th day, and by that time he had shot at 100,010 blocks. He had missed six. His longest run without a miss was 32,860, and another time he chalked up 13,017 before a block went unhit. Quite significantly, there was not a single misfire during this scorching marathron.

Ad Toepperwein, who had quit shooting by that time, promptly wrote Tom Frye and congratulated him on his remarkable effort. Toepperwein died three years later in 1962 at age 93.

Clockwork Revolvers

W. H. J. Chamberlain & A. W. Taylerson

A coiled strip of thin, *flat*, spring-tempered steel has been a power source during 400 years of time-keeping. And, as Lewis Winant's *Firearms Curiosa* shows, 17th century alarm guns, 18th century flint gunlocks, and 19th century percussion locks or bird-scarers so powered are known. We even traced a New Yorker's 1886 automatic-rifle design powered in this fashion.

However, certain firearms used such springs (referred to as "volutes" by some patentees) in the actual manner of clocks, and that design feature (which survives today) gained our attention in revolver arms. We tagged as "clockwork" any such guns with a gear train for winding their springs and then transmitting the energy so stored, and also with an escapement to measure

out that power into loading and firing mechanisms.

We also determined a "revolver" to be any hand-held firearm distinguished by a chambered ammunition-carrier (mechanically rotated on a longitudinal axis behind a single barrel) from which charges or cartridges were successively discharged, *in situ*, through the barrel. Such a firearm would also be a "clockwork revolver" if it met these two criteria. A "clockwork revolver" so defined eschews various clockwork radial or multi-barrelled and hand-turned revolving firearms (with Degtyarev and Lewis machine guns, Shpagin, Schmeisser, and Thompson subma-

This article first appeared in American Rifleman

Top: Collier used spring (missing) to turn his cylinder. (Photo courtesy of Christie's). Bottom: Edwards pill-primed revolver, like Collier's, had cylinder that rotated by spring power. (Smithsonian Photo).

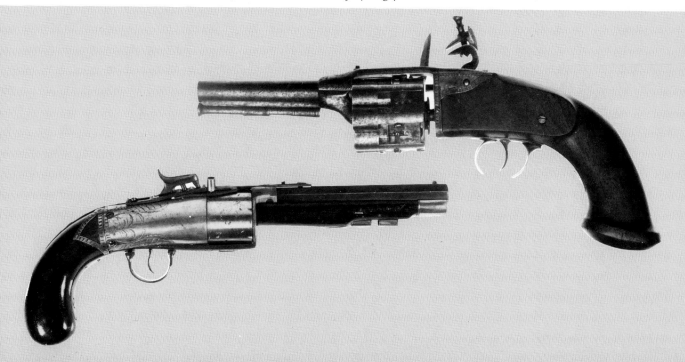

chine guns, or bolt-action rifles with Schoen-hauer and Schulhof magazines, all of which use flat spiral springs), but so we draw the lines of our concern here.

Elisha Heydon Collier's design was the earliest one to meet these tests'[1], and our picture of a rather battered English-made flintlock handgun, on his system, was chosen as showing an important part of the patented escapement.

Collier proposed a spring encircling the axis of his chambers. This spring was to be wound up by manually rotating the chambers (in a reverse direction to that in which they revolved on firing), and a helical spring then pushed the chambers forward, on their axis, to mate one of them with a male joint on the barrel to prevent rotation until required.

In theory, cocking a Collier revolver pulled the chambers back against the helical spring via a claw on the cock engaging a notched circular plate attached to the rear end of them and visible in our illustration. Thus clear of the barrel breech, the spring was to spin the chambers until the claw aligned with a notch in the circular plate. Then that helical spring was free to snap the next chamber forward onto the end of the barrel, and a ram attached to the cock secured the chambers on firing.

This design was also patented in France[2] and was originally conceived by Capt. Artemas Wheeler of Concord, Massachusetts (born 1781; died 1854), whose flintlock pepperbox carbine and single-barreled musket[3] briefly concerned the U.S. Board of Navy Commission during 1821.[4]

What novelties Wheeler claimed in his U.S. Patent, and whether Coolidge or Collier actually used flat spiral springs in "clockwork revolvers," have been disputed in the past.

We happen to think correct Willard C. Cousins' view[5] that Wheeler arms were originally intended for this type of spring, but that he abandoned it before the U.S. Board of Navy Commission. We also think that Collier often abandoned its use in the same cause of reliability.

Those holding the same view could long point only to Collier's own claims[6] to have used the spring system in some of those five- and seven-shot pistols, rifles and shotguns (percussion as well as flintlock) that the British gun trade manufactured for him, as authority for their opinion. It was Clay P. Bedford[7] who recognized (in the Nebraska State Historical Museum) a Collier flintlock shotgun still embodying both flat spiral and helical springs, as well as noting evidence that another shotgun (in the Tower of London) had once contained both features. The arguments ended.

Our next clockwork revolver appeared from Zanesville, Ohio, during 1839,[8] and we illustrate what may be the only surviving specimen of it. In this Edwards firearm the rotating spring was wound in the same manner as Collier's, but a toothed "rag wheel" on the rear face of the chambers indexed each in turn with the barrel. A spring-loaded bolt for the revolver's chambers was released by a front peg that was visible on the hammer after discharge.

Collier was still in England[9] when our next clockwork revolver surfaced there, as patented for ". . . a certain foreigner residing abroad," during 1841.[10] Various constructions of clockwork revolvers were described and illustrated in this Patent Specification; in each a flat spiral spring wound by key (or Collier fashion) powered both rotation and firing mechanisms. Thus, one version struck its percussion nipples successively

*See page 98 for numbered footnotes.

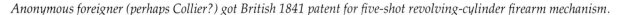

Anonymous foreigner (perhaps Collier?) got British 1841 patent for five-shot revolving-cylinder firearm mechanism.

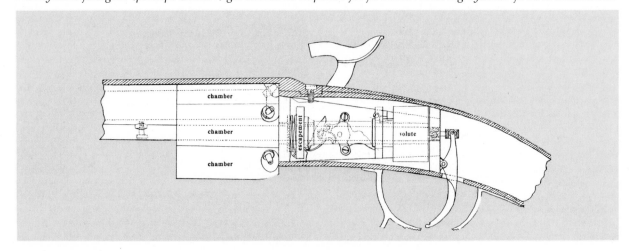

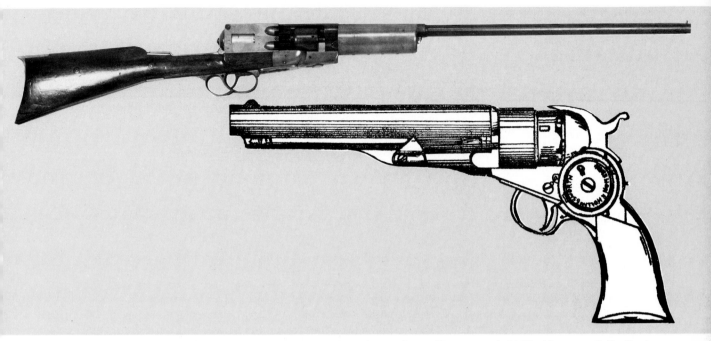

Top: Escapement regulated rotation of Hollingsworth and Mershon rifle patented 1855. (Stewart Collection). Bottom: Mershon and Hollingsworth's efforts created single-action revolver with option of self-cocking operation.

against a movable stop, as the chambers revolved; another had an external hammer cocked by pulling the trigger; and a third used a flat spiral spring to raise the hammer and turn the chambers. In the last case, a second spring released the hammer at firing.

It is disappointing that the inventor of this 1841 design—modestly publicized,[11] and our first clockwork revolver capable of burstfire—remains anonymous.

We are indebted to Henry M. Stewart for drawing our attention to the next of our clockwork revolvers and for illustrating it with an arm from his own collection. Hollingsworth and Mershon, the inventors and patentees,[12] were also from Zanesville, Ohio.

In Hollingsworth and Mershon's U.S. Patent of 1855, a "spring box" at the rear of the weapon bore the spring on an axle also bearing a toothed 'scape wheel (like that in a watch) just behind the chambers. The latter received step-by-step rotation from this spring under the control of an anchor-shaped pair of pallets rocking to intercept alternate teeth of the mechanism's 'scape wheel.

Both British and U.S. patent specifications are prolix, but we did observe that our flat spiral spring also operated the hammer in concert with rotating the chambers, to fire two or more shots without rewinding. The spring in a pistol version was intended to be wound by turning the butt.

Nearly a decade later, Mershon and Hollings-

worth launched an entirely different clockwork revolver.[13] The design involved modifying a single-action weapon to permit its user, at will, to continue thumb-cocked deliberate fire, or to change his shooting to a self-cocking mode requiring no more than alternately pulling and releasing the trigger. For the first time, these patentees envisaged using metallic cartridges.

As we illustrate, Mershon & Hollingsworth's 1863 design was to use a conventional Colt-type, single-action revolver. The conventional leaf mainspring was retained, but in combination with a flat spiral spring and a six-tooth 'scape wheel mounted on the hammer pivot. The spiral spring (wound by a folding external key) had to override the leaf mainspring, upon releasing the trigger in self-cocking function, and throw the hammer back to full-cock. Understandably, some part of the patent specification was devoted to readers skeptical of this accomplishment.

More or less coincidentally with those two Americans, one Andrew Wyley (a geological surveyor) had been developing a clockwork revolver in Staffordshire, England, as part of a particular idea for using interrupted screw threads as a breechlock in a variety of small arms employing metallic cartridges.[14]

Wyley proposed applications of his system to a single-action, gas-seal revolver, in which a spiral spring was to be wound by a quarter-turn of the butt. This act also drew the chambers back

Wyley's spectacular single-action mechanism for gas-seal cartridge mechanism used interrupted screw-thread breech.

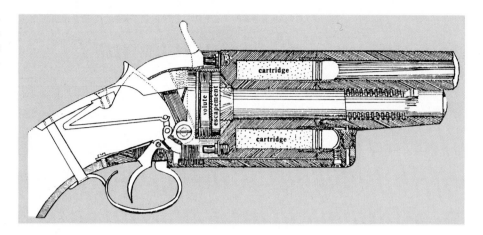

off their gas seal, by a screw on the forward end of their arbor, for rotation by the spring.

So far as we know, George C. Bunsen's was the next of our clockwork revolvers, and the relevant U.S. Patent application was filed on December 26, 1865.[15]

The Bunsen designs gave each user an option either automatically to discharge all chambers consecutively, leaving the firer to concentrate on his aim, or to fire conventionally. Since a 15-chamber waterfowl rifle for 0-size shot (each chamber double-shotted) is described in the patent specification as to be emptied in about two seconds, the idea of aimed fire seems mere puff. However, apart from the U.S. Patent model (complete with patented sliding bayonet), three production arms have survived to remind us of those men who thought they could beat the system.

To summarize all of the features of Bunsen's design is beyond us here, but some highlights of it are worth mentioning. The substantial clockwork mechanism and its spiral spring filled the butt of each weapon and operated so fast that the inventor specifically mentoned, in his patent claims, a special gear train and fan wheel solely intended to ensure the hammer hitting each cap before the cocking mechanism snatched the former back! Both hair and standard triggers were to be fitted, and the U.S. Patent specification

shyly accepts removal of the chambers to a handy pocket as the best way to prevent accidental discharge.

Mention of Sylvester H. Roper's repeating arms[16] may seem inappropriate here, since his rifles and shotguns removed each cartridge from the rotary magazine for discharge in the barrel. However, the Roper Sporting Arms Co. of Hartford, Connecticut, made some four-shot, hand-rotated rifle/shotguns that discharge each cartridge in their chambered magazine, and the eagle eye of Henry M. Stewart has noted that one such arm uses a trigger-wound spring for the rotation of its magazine.

We do not know if that spring is flat, spiral, or otherwise in form, and we do not know the source of such a novelty. It is not suggested by Roper's own U.S. Patent, and features of the earlier Draeger gun[17] could have made it difficult for Roper to patent such an improvement or to commercialize it, even if the idea was his own.

We traced no other relevant arms before 1901, when two U.S. Patents[18] were secured by Frederick B. Pope of Augusta, Georgia; both were partly assigned to Giles D. Mims of Parksville, South Carolina, and both claimed clockwork revolvers that were capable of semi- or full-automatic fire, at choice.

Pope asserted that his inventions were applicable to any revolving firearm, pistol, shotgun,

Clockwork spring mechanism, hidden inside buttstock, powered George Bunsen's revolver rifle. (Smithsonian)

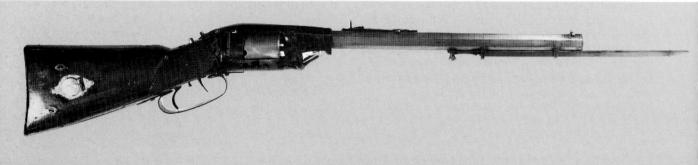

rapid-fire gun, or cannon, but our illustration (from the first patent) establishes that small-caliber, hinged-frame pocket handguns were the focus of his attention.

In such arms, the spring powered a rotating disk held, at rest, by the trigger. The disk, by pins protruding from its periphery or side-face moved the hammer back against the mainspring *seriatim*, when spinning, and continued to do so while the trigger was pressed. For obvious reasons, the design embodied a grip safety, and the second U.S. Patent covered means of winding the spring when the weapon was broken open for loading. Here a gear wheel on one of the hammer trunnions was engaged by a toothed and bifurcated "pitman" linked to the barrel lug.

At a casual glance, Pope,s design may appear no more than an interesting adaptation of a pinwheel from the strike function of a Seth Thomas or Terry clock. However, his patent specification reveals a sophisticated problem, duly answered, whereby the disk is prevented from beginning to retract the hammer prematurely in burst fire. Given a small disk, but a spiral spring powerful enough both to overcome the revolver's conventional mainspring and rotate its cylinder, such problems clearly must have existed for this inventor as they had for Bunsen.

The one seminal clockwork revolver first appeared during 1934, as a semi-automatic designed by Charles J. Manville in Indianapolis, Indiana, and it was energetically promoted (until 1941) by his "Manville Manufacturing Corporation" at Pontiac, Michigan, in 12-gauge (24-shot) and 25mm (18-shot) versions.[19] These cast-aluminum wonders were police arms, with pistol grips fore and aft like a Thompson SMG, and they fired flare, "demonstration," and gas cartridges; the small guns also handled shot or slugs. Before use, the spring had to be wound

by locking the hammer and manually rotating the chambers counterclockwise, but thereafter that power source cycled both firing and rotation mechanisms at such successive trigger pull: 24 gas shells from a 12-gauge Manville cleared its 9½-inch barrel in four seconds and yielded around 240,000 cubic feet of gas.

A Swiss/German team patented a design of concern here during 1959,[20] and the handgun application was taken up (in France) by the Manufacture des Machines du Haut-Rhin S. A., at Mulhouse-Bourtzwiller, and promoted for an uncertain period between 1962 and 1971. This Manurhin version was in 4mm or .22 rimfire calibers, with interchangeable cylinders and barrel units separately available, and it is briefly described in the June, 1974, *American Rifleman*.

The inventors envisaged their system as applicable to any kind of small arm, the users of which would benefit from rapid-fire practice using cheap, virtually recoilless, small-caliber cartridges. Thus, ammunition fed by the cylinder or by belt or by tape was applicable to their system.

A flat spiral spring powered the motor driving both lock mechanism and chambers in the version shown, and was wound by pulling out and turning a knurled cap (to be seen in our illustration) at the rear of the weapon's receiver. We do not know if this system was adapted or refined for use as a unit inserted into or mounted upon full-size (perhaps crew-served) military automatic arms, but that seems to us a more logical vehicle for it than the handgun that was actually chosen by the manufacturer.

Finally, *Jane's Infantry Weapons 1985-1986* drew our attention to a 1983 revival of Manville's clockwork revolver,[19] as the MM-1 multi-round projectile launcher of Hawk Engineering Inc., Northfield, Illinois. This embodiment (in calibers 37, 38, or 40mm, adaptable also to 12- or 28-gauge

Pope's hinged-frame revolver became automatic if its trigger was held rearward. Grip safety was featured.

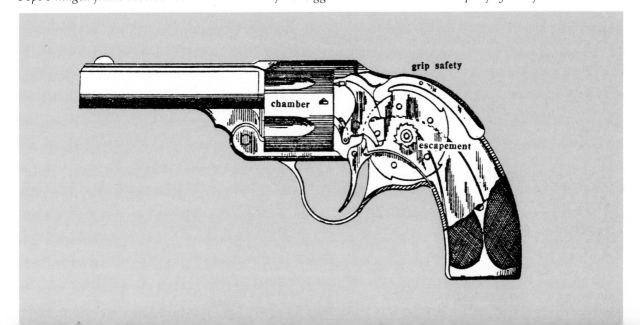

Swiss/German small-arms development in 1959 revived volute spring mechanism.

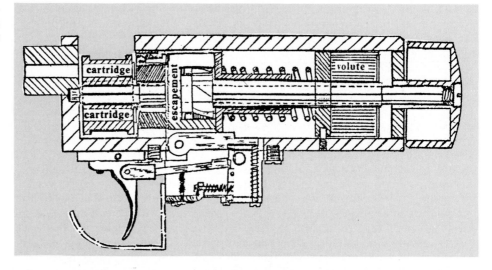

shotgun shells) loads up to 12 rounds at a time of smoke, tear gas, flare, or baton cartridges 5½ inches long.

The future of clockwork revolvers appears now to be in heavy-caliber small arms of the Manville type, but, unfortunately, detailed information about their rotation systems is still largely classified.

We know that the British Royal Small Arms Factory's 1985 ARWEN 37 Anti-Riot Weapon is not a clockwork revolver by our definition, since the feed sprocket rotation is by coiled-wire (not flat spiral) spring. However, the form of rotation

spring in the 12-gauge Striker clockwork shotgun launched in 1983 by Armsel of Johannesburg, or the South African Defense Force 1985 six-shot 40mm Armscor grenade launcher (which taps off combustion gas to unlock the cylinder for rotation), is still to be established.

All in all, it appears safe to remark that a design road sign-posted by Capt. Artemas Wheeler back in 1818 is no cul-de-sac yet.

Our grateful thanks go to those donors' names against each illustration, the authors noted below, and to Henry M. Stewart, Jr., H. Woodend, and M. J. Sarche for more general assitance.

FOOTNOTES

1. E. H. Collier: English Pat. No. 4315 of 24 November 1818.
2. C. Coolidge: French Pat. of 16 June 1819.
3. The technical specification to Wheeler's U.S. Pat. of 10 June 1818 was lost in U.S. Patent Office fire of 1836.
4. "Capt. Wheeler's Revolving Guns," B. R. Lewis in *American Rifleman*, April, 1953.
5. "Capt. Artemas Wheeler, Gunsmith, Concord, Massachusetts, 1781-1845," Willard C. Cousins in *The Gun Report*, April, 1978.
6. *Samuel Colt vs. Massachusetts Arms Co. (1851), Report of* (Boston 1852), p. 125.
7. "Collier and His Revolver," C. P. Bedford in *American Society of Arms Collectors Bulletin N. 24,* 1971.
8. D. Edwards: U.S. Pat. No. 1,134 of 25 April 1839.
9. *On the Superior Advantages of the Patent Improved Steam-Boilers Invented by Elisha Heydon Collier* (London, 1836).
10. M. Poole (a communication): English Pat. No. 9119 of 14 October 1841.
11. *Repertory of Arts (New Series)*, Vol. 18, p. 4: *Mechanics' Magazine*, Vol. 36, p. 334.
12. J. Hollingsworth & R. S. Mershon: U.S. Pat. No. 12,470 of 27 February 1885. W. E. Newton (a communication): Brithish Pat. No. 1684 of 1 August 1854.
13. R. S. Mershon & J. Hollingsworth: U.S. Pat. No. 39,825 of 8 September 1863.
14. A. Syley: British Pat. No. 1785 of 16 July 1864.
15. G. C. Gunsen: U.S. Pat. No. 51,690 of 26 December 1865.
16. S. H. Roper: U.S. Pat. No. 53,881 of 10 April 1866.
17. C. Draeger: U.S. Pat. No. 34,922 of 8 April 1862.
18. L. H. Dyer: U.S. Pat. No 666,476 of 22 January 1901—assigned to F. B. Pope and G. D. Mims. F. B. Pope: U.S. Pat. No. 666,555 of 22 January 1901.
19. C. J. Manville: U.S. Pat. Nos. 2,101,148 (of 7 December 1937), 2,151,512 (of 21 March 1939).
20. H. Vogler & H. Krausser: British Pat. No. 894,940 of 24 October 1959.

Firearms Fakery

H. Lee Munson

Faked firearms exist. They fool experts and novices alike, and when you are the victim you feel angry, foolish, embarassed, and frustrated. The financial loss can be devastating. We are often ashamed to admit our mistake to our collector friends. Sometimes we don't even admit it to ourselves. But it should not be that way. If we share our experiences, we will reduce the chances of being fooled again.

The term "fake" is difficult to define. In the strictest sense, it could mean any arm that is not absolutely original in every detail. At the other extreme, it could mean only a completely new weapon made up specifically to deceive the buyer. In most cases, however, it is neither. Some examples were spuriously marked contemporary with manufacture and use. Good examples of this are the small pocket pistols made in the 1860s marked Berringer or Derringer. These are as collectible today as original Deringers.

The classic example of "antique" arms made to delude the public is the famous set of "Paterson-Walker Colts" which were built in the 1920s and deceived the experts so completely that they became part of several famous collections and were appraised at the phenomenal price of $15,000 in 1936! They were finally exposed as fakes in 1959. The last anyone heard, they were listed in Norm Flayderman's 1973 catalog at $2,750—the set of four—as examples of faked guns. That catalog still sits on my desk as a vivid reminder that even the best experts can be fooled. Overconfidence can be as big an enemy as lack of knowledge.

Most fakes are not as elaborately conceived as the Paterson-Walkers, however. In most cases, since the faker is trying to make more than a legitimate profit on the sale, he also tries to hold his expenses to a minimum. Furthermore, few of us need concern ourselves with the complex fak-

ing of an ultra-expensive, museum-quality piece for the single reason we will never be able to afford one! What we need to be aware of are the subtle ways in which the value of an ordinary arm can be enhanced by altering details.

The detection of any faked or spurious firearm is difficult at best, but it is often made more difficult by the circumstances under which it is purchased. Gun shows are often very crowded and not well enough lighted for critical examination. Viewing time at auctions can be limited, and mail-order purchases are often made totally blind.

This article first appeared in American Rifleman

The first step in reducing the chance of purchasing a fake is to do business only with reputable dealers and auctioneers. That simple precaution can prevent many problems before they become costly experiences. Another preliminary precaution is to specialize in and know a subject thoroughly. Obviously, one person cannot have deep, thorough knowledge of 14th or 15th century armor, Colt revolvers, English flintlocks, Civil War carbines, artillery, and gambler's guns. Choose a field of interest, and learn as much about it as possible. This knowledge comes slowly, but it is the best defense against fraud.

Also under the broad umbrella of education, study as many legitimate examples in your field of special interest as possible, not only those you hope to purchase, but those belonging to friends, those illustrated in books and magazines, and those in museum collections. Learn to observe how they have aged. Note how the wood has mellowed, how wear patterns are in places consistent with use. Note the patina that develops with time on iron, steel, and brass.

Discover the subtleties of style and design that distinguish authentic pieces. Inspect the type of engraving, the depth, pattern, and placement of stampings. Notice the difference between modern, machine-cut screws and antique hand-made ones. Observe how pitting obscures original engraving, but, on re-engraved pieces, the engraving cuts through the pitting.

But enough of generalities—what about specifics? The aging process of firearms is composed of many factors, not the least of which is use. Wear patterns should be consistent with that use, and on any given specimen the wear on one part should be matched by equal wear on other parts. **Fig. 1** shows the well-worn frizzen and pan of a vintage 1790 fowling piece. If the checkering on the wrist were sharp and crisp, not worn as in **Fig. 2,** the gun would be suspect. Consistency is the key. The stock of the blunderbuss pictured in **Fig. 3** shows how the mellowing of the color the wood and its finish is affected by use. Notice the lighter area at the wrist where natural use has removed the finish.

Fig. 4 shows four screws. Two are old; two are new. Note, on the two at the left, the difference in threads-per-inch. The one at the extreme left is modern 10-32 machine-screw thread. The one on the right is not. The wood screws differ markedly in pitch, head shape, and point, the one on the right being from an antique, the one on the left, new. The width of the slot is also an indication of age. Old screws generally have much narrower slots than new ones.

The inspiration that produced this article came with the purchase of an example which embodies nearly every wrong element I've discussed. The

Fig. 1: Pitting in pan and around breech mark this 18th century piece as both well used and of genuine antiquity.

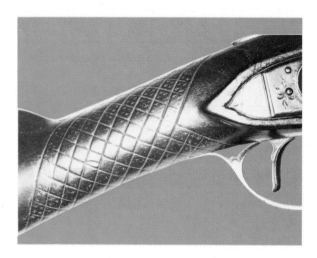

Fig. 2: Checkering can be dead giveaway in fixing age of purported antique. Diamonds, large or small, that are too sharp are probably also too new.

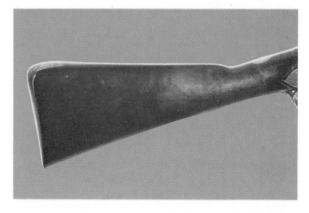

Fig. 3: Shooter's hand leaves indelible mark on old gun. It appears as mellowing of wood's natural color or change in its tone.

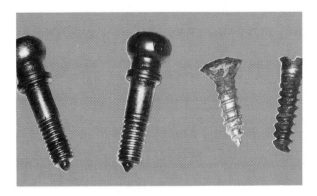

Fig. 4: Look carefully at screws. In handmade gun, standard or machine-cut thread indicates recent repair at best.

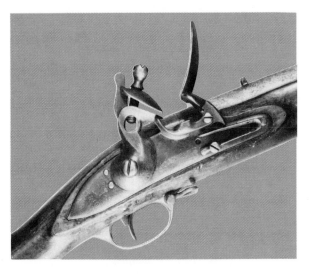

Fig. 5: Author immediately began to doubt this "Brown Bess," just because it didn't look like a Brown Bess musket.

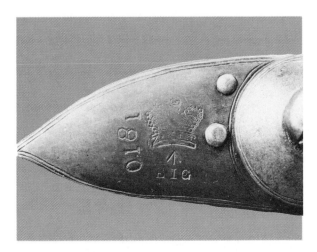

Fig. 6: E.I.C. marking is correct, but date—1810—is three years too late ever to have been on authentic lock.

subject is a British musket, bought through the mail from an advertisement in a well-known magazine published in England. The advertisement mentioned "50 Brown Bess Flintlock 39-inch muskets . . . Most have named and dated locks."

The appeal was immediate. I had been searching for years for a Brown Bess flint musket, presumably India Pattern with a 39-inch barrel, signed by Mortimer. Could it be that the advertiser had one? I wrote without delay and inquired if any of the muskets were signed by Mortimer. The reply was affirmative but added that the price was £270, not the £225 quoted in the ad.

At this state in the transaction, I had already made several critical mistakes. First, I *assumed* the dealer was reputable simply because he advertised in an internationally known magazine. I should have queried friends in England concerning his reputation and perhaps even had someone examine the musket for me. Second, I placed too great a temptation before him by requesting something specific which could be easily applied. I should have requested a very detailed description of the markings before sending the money. Third, I should have been immediately suspicious of the change in price. All caution, however, was set aside in my eagerness to acquire the musket. Thus I violated another basic tenet, which is second only to knowledge in avoiding fakes, and that is simply to approach each investment with restraint.

When the musket **(Fig. 5)** arrived, evidence of fraud was instantly visible. At first I did not want to believe my instincts, and for several days I tried to rationalize the discrepancies I saw. No one wants to admit they've been taken for a fool. But I could not reconcile the date stamped on the lock with the known dates of sales to the East India Company by the Mortimers. The last payment was made to H.W. Mortimer by the E.I.C. on March 26, 1807. The lock on my musket, shown in **Fig. 6,** bears the date 1810. Thus, thorough prior research and knowledge of the Mortimer family of gunmakers provided clues.

My knowledge of the overall configuration of the musket and what it *should* look like was far less complete. I *did* know what a Brown Bess musket of the India Pattern should look like, and this musket definitely did not conform to that design. There are at least three excellent books available which describe in very great detail the India Pattern. The classic is *British Military Firearms 1650-1850* by Howard L. Blackmore, but even more detailed illustrations of the shape and style of the stock, lock, and furniture are found in *Red Coat and Brown Bess* by Anthony D. Darling and in *British Military Longarms 1715-1815* by DeWitt Bailey.

Even if one had never examined a legitimate Brown Bess, the illustrations in these books would make it very clear that the musket I now owned could not be called by that name. Whatever it was, it was *not* a Brown Bess.

Obviously, the wording of the advertisement had been deliberately deceptive by calling the musket a Brown Bess, but since I had never seen a Mortimer–East India Company musket, I could not be sure that they followed the India Pattern. Perhaps this *was* the correct design? Inquiries were made of several experts in the field of British military muskets. The consensus was that while the musket was not a Brown Bess, neither was it newly manufactured. It was probably assembled sometime in the wide historical period between 1830 and 1870 for use in the Indian States. Thus it is a conglomerate of older and newer parts, not made up to deceive a modern buyer but to meet some long-past military need.

If the musket itself is not spurious, even though it is not a Brown Bess, how can it be called a fake? It is here that the observation of patina, natural wear in natural places, style of engraving, and the historical correctness of mark-ings becomes critically important. Patination is the only member of this list which is difficult, if not impossible, to show photographically. The nuances of color and texture can be so subtle that only first-hand observation is adequate. In this case, the lock plate is slightly more brown around the name "Mortimer" than on other areas. Wear patterns are much more photogenic. In **Fig. 7,** note how the small crown-over-3, just below the pan, is very worn, yet the letter "E" in the name Mortimer is crisp and clear. Not more than 1/16-inch separates these two marks, and if they had been there for the same length of time, they would have worn in a like manner. It is not merely that the crowned 3 was a shallow stamping, either; the edges are too smooth.

The style of engraving of the name "Mortimer" is another indication that it is spurious. I have been privileged to examine hundreds of authentically signed Mortimer arms, and not one was engraved in this manner. The method of forming the letters with three fine lines to create the illusion of width in the vertical members was never employed. An authentic signature on a genuine East India Company–Mortimer lock is

STOPPING FAKERY

Gun collectors and dealers are going to have to stop ignoring the problem of fakes in the vain hope that it will disappear. The only realistic way to stop fakes is to make it unprofitable to make and sell them. If you have the slightest question about a gun you are buying or the person you are buying it from, get a signed and dated bill of sale that states, very clearly, what it is you are buying, its condition, what is and is not original about it, plus, of course, its price. If the vendor won't sign such a bill of sale, be advised that you could be warming up for problems.

Unless you are truly expert in your collecting field, when you buy an expensive gun you would be very well advised to get a written appraisal from an expert before you spend your money. I realize that this is drumming up business for myself and other appraisers, but it is sound advice. A few hundred dollars spent now may save many thousands later. A bill of sale stating that the gun is real when it turns out to be a fake may be the basis for getting your money back, but you still have to collect it, and that can be a problem.

When you get a paid professional appraisal, you also are buying insurance on your purchase. The paid appraiser is legally liable not only for the amount spent on the appraisal, but for the value of the object appraised. If the appraiser says the gun is all original and you buy it on the strength of his written opinion, the appraiser is fully as liable to you for your money spent on the item as is the vendor.

Right now I am doing an appraisal on a fake Collier revolver. The buyer asked four friends about it before he bought it, and they all thought it "looked all right." A hard $9,600 lesson, and the buyer is still spending $300 to get the bad news in writing before the matter is given over to his lawyer, whose fees will not be low.

Just about everything worth money is being faked today. The old "favorites," of course, are reconversions to flint, refinishing, and all types of engraving. The simple fact is that it is easy to get a letter on a real engraved Colt SA, and then just "do" a plain gun, including changing the numbers, to match the letter. When buying any gun that is supported by a factory letter, remember that the letter merely states that the factory produced a gun with that serial number in a certain configuration. It says nothing about the gun you are buying! Many rare and valuable antique guns are being reproduced today: Collier revolvers, Forsyth lock pistols, flintlocks, percussion oddity and combination guns, all-metal Scotch pistols, Colt Paterson and Walker revolvers, just to name a few. Then marked reproduction arms such as Henry rifles and Confederate Griswold and Gunnison revolvers are being altered by removing the manufacturer's name and aging them. The "cautions" list is sadly without end. Be careful, and if you get stuck, holler very loudly to embarrass the seller as much as possible. Maybe he will quit! If nothing else, your fellow collectors will know that this is a man to avoid.—ERIC VAULE

Fig. 7: This "Mortimer" signature betrays itself in several ways as spurious. Style and cut of letters are incorrect; and the engraving is not worn as other marks on lock.

Fig. 8: In real Mortimer signature, letters run vertically, not slanted. In addition, the engraving on this specimen shows pits and wear that match the surrounding metal.

There are degrees of fakery. This add-on Paterson barrel is one type. Walker barrel pictured above it may only have been "restored."

shown in **Fig. 8.** Without a doubt, a number of engravers were employed by the Mortimers over the years, and probably an expert could detect differences in the work of different hands. But for our purposes, it is enough to know that the style of lettering remained unchanged.

The incongruity of the Mortimer signature and the 1810 date has already been mentioned, but there is further conflicting evidence found on the barrel. It is stamped with the heart-shaped bale mark of the East India Company and the proof marks of London, but it is *not* signed by Mortimer. Also, to be historically correct, the bale mark should not appear with a lock dated 1810, because in that year the company changed its mark to a standing (rampant) lion symbol.

Thus we have seen how small bits of evidence can be used to damn a fake. Depending on the value of the arm being examined, the equipment used in gathering this evidence can be either relatively inexpensive or extremely costly. If one is contemplating the purchase of a piece valued at $10,000, $20,000, or $30,000, then much more sophisticated examination than described herein is warranted. But if the investment is moderate, these simple precautions will save many headaches. A few simple tools can be very useful. A compact magnifier incorporating a light is worthwhile. Mine was free from the Champion Spark Plug Co., but a similar unit with a tape measure once was available for $4.50 from Edmund Scientific, 101 E. Gloucester Pike, Barrington, NJ 08007. A steel straight-edge rule can also be used to reveal an indentation where a prior mark has been removed, not to mention making measurements. A simple penlight can be very helpful for side-lighting to reveal flaws. Also useful are pocketable bore lights, tape measures, and bore gauges that can confirm measurements and uncover internal wear. None of these will help, of course, unless you remember to take them along.

One of the best reasons for avoiding the purchase of a fake is that recouping your money later is often extremely difficult and may prove too costly to be practical. In the case of this spuriously marked musket, it took over a year and even then only part of my expense was recovered. Had it not been for the Trading Standards Dept. of the Buckinghamshire County Council, I probably would have lost everything. Generally, one cannot rely on the legal system in any case of this type. The expense of hiring a lawyer and the costs of a trial are simply too great. Furthermore, the burden of proof rests entirely on the buyer's shoulders. And proof of fraud is very difficult. So prepare yourself adequately, examine each purchase thoroughly, beware of suspicious business practices, and consult your fellow collectors.

Taking the Muzzleloading Shotgun Afield

Ron Jones

Sportsmen purchasing their first replica muzzleloading shotgun are confronted with an array of decisions on the selection of loading components and accoutrements. Most new black-powder shotgunners will be asking themselves: How do I carry and measure powder and shot in the field, which powder granulation should I use, what wad components do I need, how do I develop a safe and effective load, and is shot-size selection the same for black-powder arms as it is for modern breechloaders?

If the tyro muzzleloading shotgunner is to be successful in his efforts to master the art of fowling, it is important that he understand the properties of load components and develop a feel for their interrelationships in producing a ballistically efficient charge. Following this, a thorough review of traditional loading techniques is indispensable—particularly from the standpoint of safety.

The composition of black powder has not changed significantly for 700 years. Seventy-five percent saltpeter, 15 percent charcoal, and 10 percent sulfur as a physical mixture has propelled projectiles in one form or another since at least the 13th century. While black-powder-era sportsmen had a wide variety of powder granulations to choose from, the quality was not always the best. Today's black-powder shooter is limited to just four powder granulations by a single American manufacturer, but the quality of this product is outstanding. GOEX brand black powder is offered in Fg (coarse), FFg (medium), FFFg (fine), and FFFFg (superfine) granulations (grade descriptions are the author's). Each of these granulations possesses unique burning characteristics which, in turn, determine the utility of the powder.

Fg powder is the coarsest granulation and has for years been the choice of shooters hunting with old originals. The granules look like undersize chunks of coal and produce the lowest breech pressure of the four granulations. In replicas of modern manufacture, Fg granulation produces a significant excess of fouling with little benefit in pattern efficiency. Most authorities now recognize that shooting an old original is, at best, a foolhardy venture. When all of the antique shotguns are properly shelved for their historical value, the use of Fg powder will be limited to 10-gauge and larger shotguns.

FFg granulation black-powder is the overwhelming choice of black-powder shotgun shooters. This granulation produces uniform patterns at workable breech pressures. It leaves an acceptable amount of fouling and is easily ignited by both percussion and flint ignition systems. Some shooters and one shotgun manufacturer have recommended the finer FFFg granulation for 12-gauge shotgun loads, but to my knowledge, breech pressures, velocities, and pattern efficiencies have not been published for this powder.

W. W. Greener, the 19th-century English gunmaker, was quite explicit on this subject: "Small-grained powders, whilst given (sic) greater velocity, generally cause the pellets to scatter much more rapidly than large-grained powders. The theory for this is that the finer powder burning more quickly has expended all its force before driving the shot as far as the muzzle; whilst the larger grain caused the shot to increase its velocity right up to the muzzle of the gun." To date, no one has been able to demonstrate any significant benefit from using FFFg powder in shotguns.

FFFFg granulation powder is ultra-fine grained, and its use is limited to priming the pan

This article first appeared in American Rifleman

in a flintlock shotgun. This powder should *never* be used for the main charge. For those shooters preferring to use a black-powder substitute, Hodgdon Powder Co. offers Pyrodex RS for the shotgunner. Pyrodex produces the familiar "white cloud" of smoke with considerably less residue than charcoal-sulfur-saltpeter combinations. Equivalent loads are lighter in weight than black-powder but can be loaded using volumes identical to standard FFg charges. Pyrodex is harder to ignite than black-powder, requiring hot caps and hot nipples for consistent ignition. Pyrodex usually will not ignite properly in flintlocks.

Wadding is another consideration. According to V. M. Starr, the father of modern black-powder shotgunning, the only wad you need in a shotgun is a cardboard wad of about 3/32-inch thickness. He cut his from poster board and used two between powder and shot and one over the shot. Starr resurrected the art of black-powder shotgunning more than 50 years ago and gained a reputation for developing loads in his "Sue Betsy" which would wipe the eyes of duck hunters who used breechloading smokeless-powder guns.

Actually, Starr's simplified loading concept was not a breakthrough for black-powder shotgunning. To substantiate this, we have but to go back to England in 1838 and review the teachings of England's famed 19th century sportsman, experimenter, and author, Col. Peter Hawker. On the subject of wadding, he was very succinct: ". . . perhaps the best punched wadding is pasteboard."

It is not surprising that today's black-powder shooter wants to incorporate a filler (cushion) wad into the load. Most of us still remember the rollcrimp paper loads that grandpa put together in the basement. We were told authoritatively that the fiber cushion wad was essential to the production of uniform, long-range patterns. The cushion wad was supposed to take up the slack from the initial setback when the powder ignited, thus preserving the spherical form of the pellets. Well, the concept "may" have had some validity in grandpa's rollcrimp shotshell, but I'm here to tell you that "it ain't so" in a muzzleloader! To paraphrase Starr, you can pack in fiber wads if it makes you feel better, but they won't put one extra pellet into your pattern.

But what about the convenience of modern plastic wads? They would seem to simplify the loading process immediately, and the shot cup surely will develop better patterns. I can't deny that improved pattern performance will probably result from using plastic shot cups in a Cylinder-bore muzzleloading shotgun, but the benefits are not as great as some would proclaim. You won't

get modified patterns from a Cylinder bore. Jeff Harrell, shotgun editor for *Muzzle Blasts*, gained only 4 percent when substituting a plastic wad for conventional nitro cards in a 10-shot test from a 20-gauge percussion gun in 1981. Comparable improvement can be had by using high antimony shot in standard card-wad loads. Plastic wads were not designed to obturate at the low breech pressures generated by muzzleloading shotguns. As a result, gas leaks around the base and reduces the already minimum velocity by as much as 100 fps. As we'll see later, frontloading shotgunners cannot afford to sacrifice muzzle velocity

below that generated by equal volume loads using card wads.

It's safe to assume that the majority of sportsmen taking up the black-powder shotgun are familiar with shot-size and shot-volume requirements for various game species. Most will have hunted with modern arms and will have acquired a feel for the shot sizes required to grass a pheasant, duck, or goose. Unfortunately, the information gained from years of experience with a modern firearm is not uniformly applicable to the muzzleloading shotgun.

Frontloading scattergun muzzle velocities are

1 2 3 4

Shotgun loading sequence begins with ramrod withdrawn and gun braced between shooter's legs (Fig. 1). Black-powder charge is poured from flask into measure (Fig. 2), never directly into barrel. Making certain gun muzzle is angled safely away from his face, loader charges bores (Fig. 3). Two nitro card wads seated over powder charge (Fig. 4) help produce good patterns, advises author.

5 6 7 8

Shot, measured equal volume to powder charge, is metered into each barrel (Fig. 5). After lubing muzzle, author splits wad with thumbnail (Fig. 6) and seats half in each barrel. Gun-mounted ramrod will suffice in field, but author recommends sturdier bench rod (Fig. 7). Black-powder shooter Marv Lear caps 12-gauge CVA double (Fig. 8), last step in preparing to fire.

TABLE I				SHOT SIZE SELECTION		
	Modern Shotshell Ballistics				Equivalent Equal-Volume* Black-Powder Loads	
Distance (Yds.)	Shot Size	Velocity (fps)	Downrange Energy (ft.-lbs.)	Shot Size	Velocity (fps)	Downrange Energy (ft.-lbs.)
20	9	1150	1.07	9	1050	0.95
20	8	1150	1.63	8	1050	1.4
30	7½	1150	1.5	7½	1050	1.3
35	6	1200	2.5	5	1050	3.0
40	6	1330	2.5	5	1050	2.7
40	5	1330	3.6	4	1050	3.6
40	4	1330	4.8	3	1050	4.5
40	3	1330	6.2	2	1050	6.3
40	2	1330	8	BB	1050	12.4

*See text for explanation of equal-volume loads.

nearly 300 fps slower than modern high-velocity loads, thus producing significantly less energy per pellet than modern loads. Actually, the logic we must follow is not unlike that deduced by those who developed steel shot loads for breech-loading shotguns. Since pellet energy is a product of velocity and mass, the steel-shot people beefed up the energy of their low-mass steel pellets by increasing the shot size (initial velocities were also increased slightly).

Muzzleloader shooters must use similar logic to bolster the energy of low-velocity loads. An easy maxim to follow is to use one shot-size larger than the "minimum" shot-size recommended by modern shotshell manufacturers. If 4s, 5s, 6s, or 7½s are recommended for pheasant, don't use shot smaller than 6s. If 4s, 5s, or 6s are recommended for ducks, choose 5s as your smallest shot-size. If you compare the energy of pellets launched at muzzleloading vs. modern velocities (Table I), the logic of this concept is evident.

It would be nice if choosing a large shot-size put our muzzleloading shotgun load back on a par with its modern smokeless-powder counterpart.

Regrettably, by increasing the pellet size, we reduce the number of pellets in our load. If 1⅛ ounces of modern 6s produces sufficient pellet density to grass a pheasant at 35 yards with a given choke, an identical load of black-powder 5s will come up short under identical conditions. The black-powder smoothbore shooter has three alternatives to put his low-velocity load back on a par with his breechloader. He can: (a) reduce this maximum range by 5 yards; (b) use one degree more choke; or (c) increase the weight of his shot by 20 percent. For most of us, reduced range is the answer.

One apparent contradiction to this rule is the selection of shot sizes for clay-target work. Popular shot sizes used by modern competitors work equally well in the black-powder clay-target sports. Skeet targets break as effectively at 21 yards with No. 9 shot, and trap targets fragment with ease with 8s.

According to historians, birdshot was initially produced by hammering or rolling lead into sheets, then chopping it into small cubes.

Obviously, the aerodynamics of this form of shot left much to be desired. Large shot sizes (buckshot) could be molded, but the shot tower and its arsenic-laden "drop shot" was the key component in making the shotgun an effective tool for wingshooting. While Manton's experiments with quicksilver (mercury) provide optimism for those wishing to improve the efficiency of relatively soft drop shot, the addition of antimony to form what is now termed chilled shot proved the most significant contribution to improved performance.

Today, the black-powder shotgunner has two types of unplated shot to choose from, chilled and high-antimony (magnum). Modern trap shooters have reported up to 13 percent improvement in the patterning efficiency of their loads by switching to the harder shot, but we must remember that their guns are subjecting the shot column to greater disruptive forces than the muzzleloader.

In order to determine whether there was any significant advantage in using magnum rather than chilled shot in muzzleloaders, I ran a comparison test on a black-powder shotgun known for its "kid gloves" handling of shot. I reasoned that if my soft-shooting Cylinder-bore 12-gauge flintlock fowler patterned better with magnum shot, then choked percussion guns would undoubtedly fare as well or better. The results were not remarkable, but they did demonstrate that even a shotgun which cradles its shot like a mother with a newborn baby will benefit in some measure from high-antimony shot. I evaluated the patterns using Oberfell-Thompson criteria. They averaged 5 percent better patterns with magnum shot, with fewer patches. The effective pattern was larger in diameter, and the core had increased in pellet density.

I feel that it is safe to deduce from this test and the tests of others that high-antimony shot increases the effective pattern size, as well as its long-range potential. The benefits will probably not be as great as those claimed by modern shooters, but it may spell the difference between a cripple and a clean kill on occasion. At 15 cents per 25 rounds, the benefit seems worth the expenditure.

Just a word on steel shot in frontloaders. The significant factor here is the inadequate pellet energy produced by the combination of low-density steel and low-velocity black-powder loads. The combination would cripple more ducks than it would harvest humanely.

Check any compendium containing black-powder 12-gauge loads, and you are sure to find reference to 4-dram, 1½-ounce duck loads or 3¼-dram, 1⅛-ounce upland loads. These loads generate velocities in the neighborhood of 1200 fps, not too different from modern breechloading shotgun velocities. For the new black-powder shooter, it makes sense to achieve velocities similar to those of modern arms. Yet visit any black-powder trap or skeet competition, and the majority of shooters will be using equal-volume shot and powder loads.

Measure an equal volume of FFg powder with a shot dipper set at 1½ ounces of shot, and you come up with approximately 3½ drams of powder. Set the dipper at 1⅛ ounces of shot, and that volume will measure 2¾ drams of powder. Place a chronograph downrange, and the velocities you measure will be in the neighborhood of 1050 fps. But isn't it reasonable to attempt to wring all of the velocity we can from the charcoal burners? Shotguns are not high-velocity firearms, and their effectiveness depends as much on the number of pellets they will place in a given area at a given distance as on the force with which each pellet hits. Looking at the winners of the London Public Gun Trials of 1859, we find that these gunmakers used equal-volume loads of 2¾ drams of powder to drive their 1¼-ounce loads. Go back to the works of Col. Hawker, and you will find that equal-volume loads proved the most effective even in the flint era (prior to 1825). And so it is that the equal-volume load pushing its payload of shot out of the muzzle at 1050 fps produces the most efficient black-powder shotgun loads you can build. Use less-than-equal volume, and you sacrifice too much velocity for all but short-range targets. Go above equal volume, and you sacrifice patterning efficiency. You also get more recoil!

Equal *volumes* (**Table II**) of shot and powder separated by cardboard wads make an extrememly efficient, yet simple field load. In a 12-gauge shotgun, 1 ounce of magnum 7½s or 8s propelled by 2½ drams (68 grains) of FFg will produce a beautifully patterning load in nearly all Cylinder-bore guns. Grouse, quail, and woodcock are excellent quarries for this load. For larger birds such as pheasant, 1¼ ounces of magnum 5s driven by 3 drams (82 grains) of FFg will do the trick out to 35-plus yards in a Modified barrel. If your gun will accept heavier loads, 1½ ounces of magnum 4s is good medicine for larger ducks in a Modified- or Full-choked barrel. Three-and-one-half drams (100 grains) of FFg will give this load sufficient velocity for good penetration at 40 to 45 yards.

All of these loads are easily constructed using one wad—the old fashioned 0.125-inch nitro card. Carry a handful in your pocket, and place two between the powder and shot and half of a third wad over the shot. With a little practice, you'll find that it is a simple matter to split a nitro card in half with your thumbnail to make two overshot wads. If you're loading a double barrel, pop one into each barrel.

There's only one additional ingredient you will need—a lubricant. The old timers just spit in the

TABLE II EQUAL-VOLUME SHOTGUN LOADS				
	Shot Wt. (ozs.)	Approx. Dr. Equiv.	FFg Wt. (grs.)	Pyrodex RS Wt. (grs.)
20-Ga.				
Target	7/8	2¼	60	48
Field	1	2½	68	54
16/14-Ga.				
Target	1	2½	68	54
Field	1⅛	2¾	75	60
12-Ga.				
Target	1⅛	2¾	75	60
Field	1¼	3	82	66
10-Ga.				
Target	1¼	3	82	66
Field	1½	3½	96	77

Approximate velocity using nitro card wad column about 1050 fps. Note: These loads not recommended for original (antique) black-powder arms. Never exceed replica arms manufacturer's load recommendations.

barrel, but I refuse to place my mouth over the barrel of a loaded gun. When authenticity borders on insanity, I refuse to mimic the old masters. Instead, I carry a small plastic squeeze bottle (empty Visine or nose-spray bottle) full of liquid Crisco or black-powder solvent and run a bead of it around the inside of the muzzle just before seating the over-shot wad.

The shooter soons learns to use just enough to coat the inside of the barrel without soaking the nitro card below. The wad distributes the lubricant along the inner surface of the barrel, keeping the fouling soft. If you don't use a lubricant, it will become exceedingly difficult, if not impossible, to seat the wads after several shots. The fouling will also destroy your patterns.

Just a word on ramrods. Those provided with most field guns are difficult to load with and can be dangerous if the wads seat hard. Unless you are after grouse or rabbits in thick cover, it is advisable to take along a bench rod whenever conditions permit. Duck, dove, and turkey blinds are ideal places to take your 1½-inch thick hickory bench rod or ball-capped steel loading rod. If something happens to it, you'll have your under-barrel rod as a backup.

You will also need to purchase a nipple wrench, vent pick, powder measure, capper, shot belt, and powder flask before heading for the field. A nipple wrench and spare nipple should be tucked away in a pocket in case a replacement is needed, and a vent pick is useful to probe a clogged nipple vent. A measure is a must for transferring powder charges from flask to barrel. A capper isn't absolutely necessary, but it is often difficult to grasp a loose cap in your pants pocket among the withered fern leaves and plant seeds—especially if the fingers are cold.

Shot is traditionally carried in a shot pouch (hard or soft) or a shot belt. Belts were more popular in the flint era, while the flask predominated in the percussion era. Both are equipped with tips which allow you to meter a predetermined volume of shot. The Irish tip has a scoop which carries the shot from the flask or belt to the barrel, while the English tip requires that you place the spout of the flask or belt over the barrel, then release the charge by squeezing the lever. In general, the shot belt seems to be better served with an Irish tip, while the flask appears handier with an English tip.

Powder can be carried in either a horn or flask, but traditionally was transported afield in copper, brass, or zinc flasks. Most were fitted with adjustable measuring heads which allowed the shotgunner to select from three or four charges. Replicas of Dixon or other fine old original flasks are available which are quite accurate in measuring powder charges.

A word of caution! These adjustable measuring heads cannot be used safely to transfer a charge of powder directly to a recently fired barrel. A latent spark in the barrel will instantly turn your flask into a grenade. Always carry an open-ended powder measure with you and use the measure to transfer the powder charge to the barrel. Thompson/Center makes an excellent adjustable measure which works well.

All of this paraphernalia may seem like a lot of extra gadgets to carry into the field, but each has a specific purpose. When game is active and the shooting is hot and heavy, you'll be glad you invested in these accoutrements.

At the trap range where choked guns are employed, most shooters choose to use a ⅜-inch filler wad in their 12-, 11-, or 10-gauge trap guns. The logic of this loading technique stems from the ease with which a solvent-soaked filler wad transmits lubricant to the barrel. Filler wads are placed in a small vessel of "moose milk," powder solvent or just plain water prior to the start of the shoot, then loaded damp (excess solvent squeezed out). Most guns will shoot "one shot clean" when loaded in this manner. It is important to squeeze excess lubricant from these wads before loading, as excess liquid adds significantly to the charge weight and undoubtedly raises breech pressures.

These wads should be used *only* in mild, well-below-maximum loads. They are seldom used in the field because it is possible for excess lubricant to saturate the nitro card and inactivate the powder main charge. Water-based lubricants may also rust the barrel when left in contact with a fouled barrel for a long period. In Cylinder-bore guns used in skeet, these wads tend to distort patterns. Presumably, these moisture-laden wads, unimpeded by a choke, momentarily overtake the shot charge as it leaves the barrel and hasten the spread of the shot column. The nitro card wad column is best here.

I'm sure you're hoping I'll tell you that the use of equal-volume loads with nitro card wads will guarantee predictable, uniform patterns in all replica arms. But you know I can't! Like their modern counterparts, each gun is a law unto itself. Twenty-five-yard patterns are useful for Cylinder-bore guns. Try for 70 to 75 percent patterns, with an exceptional load going two to three points higher. Choked arms should be patterned at 40 yards. Look for patterns of 50 percent in a 30-inch circle for Improved Cylinder chokes, 60 percent for Modified chokes, and 70 percent or better for Full-choke barrels. Once you've developed maximum efficiency in your loads, head out to the range for a few rounds of black-powder trap or skeet. The practice is essential to safe and effective use of these arms in the field.

PART FIVE

AMMUNITION, BALLISTICS, AND HANDLOADING

How to Get Started in Handloading

John Wootters

When I crammed the bullet into the first round of ammunition I ever reloaded, shortly after World War II (or was it the War of 1812? . . . the memory grows dim), reloaders were few and far between, and widely regarded with the same suspicion formerly reserved for alchemists, witches, and warlocks.

Today, handloading is the "in" thing. We handloaders are millions strong. Far from being one of the black arts, reloading is thought of as a normal, logical, and even essential route toward improving performance in firearms, as well as saving money. The current reloading data handbooks occupy a long shelf, and manufacturers burn the midnight oil dreaming up clever new gimmicks to please us.

A phenomenon common to both eras of the handloading hobby, however, has been the tendency to forget that not everybody with an interest in the subject is an expert. There are shooters out there who are interested but may be a little intimidated by the whole idea and don't quite know where to start. To such people, most of what is written on the subject is mystifying at best, gobbledegook at worst.

Beginners and would-be reloaders of the world, take heart! What follows is for *you*.

There are two basic divisions of handloading. The first is usually called "metallic reloading," referring to those cartridges whose cases are made entirely of metal. These days, "metallic" means rifle and pistol cartridges, and distinguishes them from the other category, more descriptively termed "shotshell" reloading. Although loading these two kinds of cartridges is somewhat similar, they do differ enough that they cannot be dicussed efficiently and clearly in the same short article. Therefore, the present piece will deal exclusively with the metallics.

A cartridge comprises four components—the case, primer, powder charge, and projectile. Of these four, the last three are all consumed or expended in firing, leaving only the brass case (usually it's brass, but casings will occasionally

This article first appeared in Guns & Ammo

Stanley W. Trzoniec photo

be found that are made of aluminum, steel, or some combination of metals). Fortunately, since it happens to be the most expensive component, the case is reusable, and this is the foundation of reloading. In simplest terms, the empty case is processed to restore it to correct dimensions, the spent primer is replaced with a fresh one, a new powder charge is dropped into it, and a new bullet is seated. In practice, those steps are somewhat more complicated than that description makes them sound, but not much.

The number of times a single case can be fired safely varies with several factors, but a reasonable average is probably about 10 to 12 times. With certain cartridges, case life is virtually unlimited; I have a few lots of brass (which is the handloaders' term for cartridge cases) that have endured as many as 40 firings each and remain sound. In certain other calibers, six or eight firings may represent satisfactory case life. However, it's easy to see that if we assume a new case costs a quarter, we can amortize that investment by multiple firings down to somewhere between two cents and essentially zero.

Practically speaking, reloaded cartridges cost between 1/3 and 1/2 the price of factory rounds, or even less, depending upon the type of bullet chosen. Economy, thus, is a potent incentive for reloading, although it is by no means the only one.

It is necessary, however, to offset the cost of the tooling required against the savings in ammo costs in order to have a realistic picture. The basic

With bench surface of only 17 × 22 inches, this loading setup was built into one half of closet.

minimum equipment, purchased new, can cost from about $75 up to about $250. The Cadillac-level gear can set you back $500 to $700 or even more, but about $175 is probably a reasonable average. Obviously, this figure can be reduced substantially by purchasing used equipment or when two or more shooting friends pool their resources.

What it all boils down to is that the initial investment in tooling will probably be amortized completely in two to five years, depending upon how much you shoot.

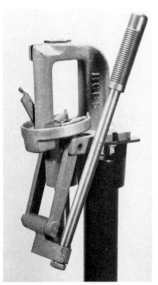

Single-stage press like RCBS' Rockchucker (left) will serve both beginners and pros quite well. Powder measure (center) is handy for speedy production and is very consistent with most types of propellants. Next step up in convenience and speed is tool such as Lyman's T-Mag turret press on Lyman's portable bench (right).

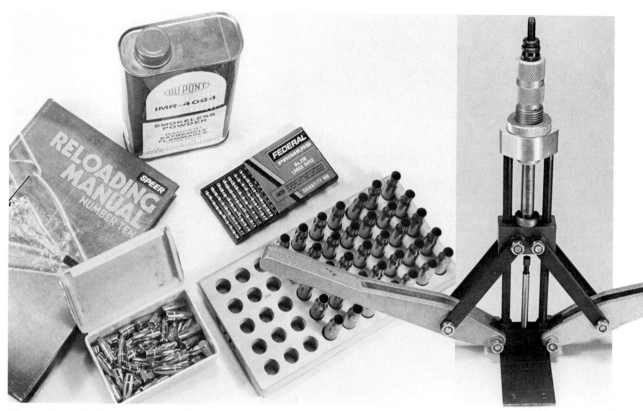

Elements needed for production of quality ammo include powder, bullets, primers, and cases, plus loading manual to guide you along (Fig. 1). Huntington's Compac (Fig. 2) is hand tool that accomplishes all loading functions and accepts standard reloading dies. Its portability complements its low price.

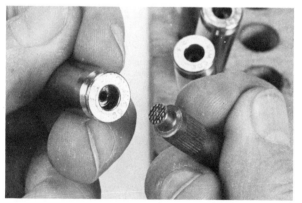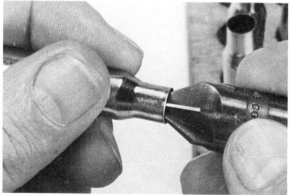

After depriming and sizing, primer pockets should be cleaned of any residue that has accumulated (Fig. 3). With new cases or freshly trimmed brass, the mouth should be dressed with good-quality case-mouth chamfering tool (Fig. 4).

As illustrated by this lineup of .30/06 brass, handloader will encounter wide variety of headstamps on cases. For safety and also efficiency, cartridge cases should be sorted by brand or lots.

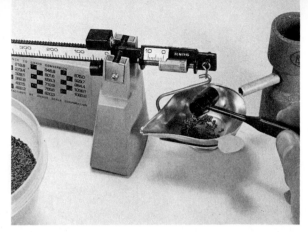

Most bench-mounted reloading tools feature priming arm, shown being positioned in ram (Fig. 5). A powder scale (Fig. 6) is essential. Loading is possible without measure, but not without good scale.

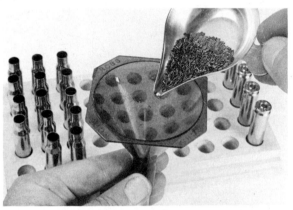

Before bullet seating, powder is carefully dumped into each primed case (Fig. 7). Funnel is helpful. Bullet is held atop case while case is moved into seating die (Fig. 8). This helps assure proper alignment.

The truth is, however, that a new reloader probably won't actually save a dime because he'll shoot so much more once he begins to develop loads for his own guns! It's more realistic to say that he will shoot his rifles and pistols from two to three times as much for the *same* cost.

If that sounds good, consider that economy is most likely the *least* important reason for handloading to most of us. Our real motivations are diverse but include improvement in the accuracy and/or power (sometimes) of our firearms; the ability to devise special-purpose, custom loadings not available from the factories; selection of the one precisely perfect bullet for a given purpose; or the need to furnish ammunition for a foreign, obsolete, or wildcat gun for which commercial ammo is not available.

Mechanically, all operations in reprocessing a fired case are performed in dies. What follows assumes a simple bench-mounted reloading press. With most of the handheld models, the order of operations will vary somewhat, but the same things will be accomplished. The first die in a conventional two-die set, which is the kind used for all bottlenecked rifle cartridges,

squeezes the case diameter back down near original factory specifications, punches out the fired primer, reduces neck diameter, and then re-expands the neck to hold a new bullet firmly.

A fresh primer is then seated in the primer pocket, using either a separate tool or attachments on the loading press itself. Next, a carefully measured powder charge is placed in the primed case. Finally, the second die in the set seats a new bullet to the proper depth and if desired, crimps it in place.

If a three-die set is in use (for straight-sided rifle and pistol cartridges), these operations are performed at different steps, but the main difference is that resizing and expanding are done in different dies. Depriming may occur in either die, depending upon the manufacturer's ideas.

That doesn't sound very complicated, and it really isn't. There are, however, a few non-mechanical steps which I left out in that brief description. Cases must always be inspected carefully before being recycled to ensure that they're sound. They may also require chamfering and deburring of the case mouth. If already fired several times or if the bullets are to be crimped in

Reloaders can obtain seemingly endless variety of bullets, including many designs of cast bullets. This variety allows you to tailor ammo for either game or targets.

place, the cases may have to be trimmed back to a specified length, after which they will need chamfering and deburring again.

Before insertion into the sizing die (unless it's a carbide die, used mostly for pistol cartridges), they must be lubricated. After resizing, the lubricant must be wiped off each one. After several firings it's best to clean the residue out of the primer pocket before repriming. For each of these steps, there are simple and relatively inexpensive tools available.

With a group of resized and primed cases in a holder, termed "loading block," we come now to the interesting part. A charge of powder must be measured and/or weighed out and dropped into each case, making certain that each case gets a full charge and *only* one charge. This absolutely requires a powder scale. Even if you have a powder measure, the charges it drops have to be adjusted and checked on a scale. You can reload good ammo for the rest of your life without a measure (if you're in no hurry), but you cannot do without a scale.

The final action is seating of a preselected bullet. Bullets for revolvers and tubular-magazine rifles must have the case mouth crimped into a cannelure, or groove, on the bullet provided for that purpose, to prevent movement of the projectile in the case in either direction under the forces of recoil.

The inevitable question at this stage of the discussion is, "But how do I know which bullet and what kind and how much powder to use?" This brings up the first and probably most essential purchase to be made by any beginning handloader, which is a reloading manual or handbook. These are published by manufacturers of bullets (Speer, Hornady, Sierra, and Nosler), powders (Hodgdon, Accurate Arms, Winchester-Western, DuPont, Hercules, and others), reloading tools (Lyman), and several special-interest are merely publishers (Digest Books, the NRA, etc.)

Many of these manuals feature a thorough dis-

cussion of the reloading process and of the publisher's reloading products, if any; some offer sections dealing with special problems and purposes; and several include extensive ballistics tables. All stress reloading safety, and all provide very extensive loading data. These data include a range of powder charges with each suitable propellant for several different bullet weights and styles in every caliber. With a little study of one (or preferably several) of these handbooks, it's no trick to pick a powder out of the 50 or more different kinds on the market today and a safe starting load for almost any type of bullet. Moreover, the product descriptions in the bulletmakers' books will at least simplify the process of selecting the correct bullet for big-game hunting, varmint shooting, target shooting, plinking, or whatever.

It will be noted that I mentioned a *starting* load. This implies the most fundamental procedure to ensure safe and satisfactory reloads in any metallic cartridge: You start out with a mild powder charge and work cautiously upward, monitoring all symptoms of pressure as you go, until your own gun tells you to stop. The process is called, simply, "working up a load." Typically, you reduce the weight of the starting load by 5 to 10 percent below the listed maximum and assemble three to five rounds with that reduced charge. A similar number of rounds loaded identically except that the powder charge is increased one grain in rifle cartridges, ½ grain or less in handguns, and another set with yet another increment of powder, and so on. As the maximum load is neared, it's prudent to decrease the increment to ½ grain in rifle cases and ¹⁄₁₀ grain in pistols.

The resulting series of cartridges is called a pressure series, since its purpose is to determine the maximum safe charge of *that* powder with *that* bullet in *that* gun. It seems to come as a surprise to many novices that no two firearms, even of the same make and caliber and with consecutive serial numbers, can be expected to produce identical chamber pressures and velocities with identical loads. Since your gun and the one used by the publisher of the handloading manual to develop his data are not the same, what is a maximum safe load in his gun may very well be a dangerous overload in yours . . . or vice versa. The same is true of your gun and your buddy's.

The one and only means available to establish maximum powder charges in your own rifle is that described above—loading a pressure series and firing those rounds, beginning with the lightest charges. In the process you must watch very carefully for *any* sign of high pressure, as described in most reloading manuals. These include, but are not limited to, excessive blast and recoil; excessive effort required in opening the action; cratered and/or flattened primers; shiny marks on the case heads; and excessive case-head expansion, as measured with a micrometer. Some of these symptoms can be produced, in certain

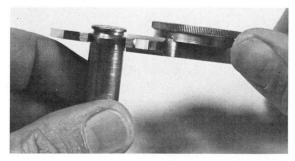

Case is being miked for signs of pressure. Excessive pressures can result from extremely small variations in charges.

Straight cases and some obsolete calibers require three dies for loading. Forming die (standing, lower left) must be used for some oddball and obsolete cartridges.

In addition to basic tools such as press, dies, and powder scale, small gadgets and accessories are of great help. They include funnel, chamfering tool, powder trickler, and primer pocket cleaner.

Bullet puller is extremely useful to shooters who develop experimental loads. This one from RCBS is efficient and easy to use.

firearms, by conditions other than high pressures, but none of them can ever be ignored until the actual cause is positively identified. Also, there are certain other indications not listed above, the determination of which requires expensive equipment.

A detailed discussion of these pressure signs and their interpretation would require another article at least as long as this one. Suffice it to say that sound, well-established, and relatively simple procedures by which even a beginning handloader can know that his loads are safe do exist. In the present state of the art, there's no way he can determine the exact pressures (in psi, copper units of pressure, or whatever) produced in his guns, but that isn't important. What is important is that there are ways in which he can be quite positive that his loads are not dangerous.

Safety is the first consideration. After that, careful benchrest shooting at different distances can establish accuracy and trajectory. In a nutshell, that's the process called "working up a load." It assures sound, effective working loads

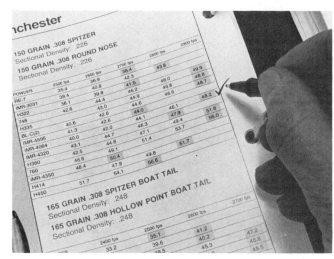

Shooting groups from bench is essential in developing and testing handloads. Always develop your loads from recognized source of data. Here, author is using manual published by Sierra Bullets.

which punish neither the gun nor the shooter unnecessarily and which get the job done at least as well as factory ammo. And it requires no Ph.D. in physics, or any elaborate equipment—just patience and common sense.

Considering that there are already between three and five million of us doing it, that reloading accidents are extraordinarily rare, and that I have known children as young as eight years old who (with parental supervision, of course) produced perfectly satisfactory ammunition, there's no reason you should have doubts about your own ability to do so.

Incidentally, a handloader will encounter no licensing or other legal ramifications (unless he sells reloaded ammunition) other than possible municipal fire-code regulations on powder storage. It should be added that modern smokeless gunpowder, correctly stored in normal handloader's quantities, is not nearly as great a hazard as gasoline, cleaning fluid, paint thinner, or many other substances that may presently be in your house or garage.

The basic handloading equipment which will be needed has mostly been mentioned already: a press, dies, and powder scale. A powder measure is a great speeder-up of the process, and there are several small, inexpensive accessories that most of us regard as indispensable as well. These include a loading block to hold cases, a powder funnel, a powder trickler (with which to add powder to the scale pan granule by granule), a chamfering/deburring tool, an inertia-type bullet puller, and a primer "flipper." A case trimmer will eventually be required. A dial-indicating caliper is an excellent investment, and you will accumulate a few screwdrivers, Allen wrenches,

and perhaps a pair of pliers and/or a small crescent wrench on the loading bench. With this equipment in hand, expanding your operations to reload for additional cartridges usually requires only the purchase of another set of dies.

Later on, with your feet firmly on the reloading ground and your interests established, you may wish to invest in an almost endless variety of specialized tooling and instrumentation, from chronographs to automated, progressive loading machines to sophisticated measuring devices or equipment for making your own bullets. Or you may not; such refinements are definitely not required, and most reloaders never feel an overpowering need for them.

A thing which is, surprisingly, seldom mentioned about reloading is that it is, in itself, a pleasant and relaxing activity, and not merely a chore whose sole purpose is to provide cheap ammo. Your typical, addicted handloader is forever scheming, studying, and plotting new ways to improve his firearms' performance, and some of us have been known to load more ammunition than we had any need for, simply because we enjoyed the process. And there is a feeling of gratification that's hard to describe when a new load produces the smallest group in a given rifle's history or puts down a pronghorn cleanly at long range with a cartridge you concocted with your own hands.

Furthermore, the term "amateur ballistician" might be substituted for "handloader." Reloaders, by the nature of their hobby, understand more about how firearms function and are generally far better marksmen than shooters who stick to commercial ammunition. That alone is almost worth the price of admission.

Time the Speeding Bullet

Jim Carmichel

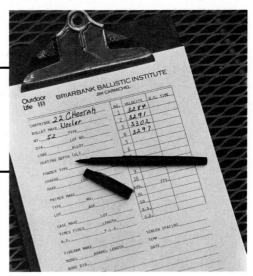

Velocities of a string of shots are recorded. Note that there are tiny shot-to-shot variations.

Any time you select a new hunting cartridge or debate the relative merits of cartridges or compare one with another, the single most important consideration has to be bullet velocity. Everything a bullet is capable of doing—or incapable of doing—depends almost entirely on its velocity. The very stuff of ballistics is bullet speed. When we know how fast a projectile leaves the muzzle of a firearm, we can predict almost everything about its flight, but if we don't know its speed, we know almost nothing about what a bullet or shot pellet will do. That's why ballisticians and hobby experimenters go to such elaborate lengths to determine how fast bullets travel. In fact, ballistic science has been closely paralleled by the development of ways to measure a bullet's speed. As these methods have improved, so has our understanding of the bullet's flight.

Because marksmen and hunters of old couldn't see their bullets in flight, they could only guess at what was happening, and it was only natural that some fascinating theories were advanced. One notion held that a bullet traveled in a straight line until it ran out of speed, at which time it simply fell to the ground.

EARLY RESEARCH

The great Isaac Newton was fascinated with ballistics, and he devised formulae that are essential in the study of a bullet's flight. He was well aware, however, that improved knowledge of ballistics would be possible only after a means of measuring bullet velocity was perfected. To do that, he spent a good deal of time trying to find a super-accurate way to measure small increments of time. Newton realized that if he could measure the interval of time required for a bullet to travel from Point A to Point B, he could easily

calculate its velocity. But no such timepiece was developed during his lifetime or for many generations thereafter. Newton's only recourse, therefore, was to make a few general speculations about a bullet's journey through the atmosphere.

He was correct in his assessment that the atmosphere plays a role in a bullet's speed, but he grossly underestimated how important that role actually is, so Newton went to his grave without unlocking the secret of bullet velocity. Ironically, he had the key all the while, but didn't know how to use it.

Newton's unused key was his own mathematical formula for determining the impact energy of a moving body. If the speed and weight of the body are known, the energy can then be precisely calculated. During the 1740s, an Englishman named Robins made the reasonable deduction that by reversing Newton's arithmetic, the velocity of a moving object could be precisely calculated. In other words, if the weight and energy of a moving bullet were known, then it would be a fairly simple matter to calculate the bullet's velocity. But how could one determine a bullet's energy without first knowing its velocity? Robins' brilliant solution was the ballistic pendulum.

Robins, like all shooters, observed that a target moves when it is hit by a bullet. He carried the observation a step further, and reasoned that if a bullet moves a target of known weight a measurable distance, it will then be possible to calculate the bullet's striking *energy*. That was the answer! Robins put his reasoning into practice by

This article first appeared in Outdoor Life

Oehler Model 33 Chronograph with Sky Screens provides handloader with immediate readout.

Air-pistol pellet provides hard test for chronograph because projectile is quite small and slow.

designing a swinging pendulum of known weight. The pendulum swung when hit by a bullet, and by measuring how far the pendulum moved, Robins was able to calculate the bullet's energy and hence its velocity as well.

As so often happens in scientific discovery, Robins and his invention were scoffed at, especially by the British army. His close-range tests indicated that a musket ball exited the muzzle at about 1,500 feet per second. At 100 yards, however, his calculations showed a remaining velocity of only about 1,000 fps. Robins explained the rapid loss of velocity as being due to air resistance, but the army experts just wouldn't buy it. After all, they pointed out, Newton himself hadn't anticipated that the atmosphere would cause so much resistance. In time, it was proven that Robins was right and Newton, for once, was mistaken. Modern tests with sophisticated equipment have confirmed Robins' velocity calculations for the old British musket.

Though the ballistic pendulum was a brilliant idea and was theoretically precise, in actual practice it was slow to operate and subject to errors in readings and arithmetic. Naturally, ballistic scientists eventually began fidgeting for a velocity-measuring tool that was faster and easier to operate. Such equipment, however, was not possible until the age of electricity, which brings us to the close of the last century.

With electricity, everything seemed possible, and the measurement of a bullet's speed was no exception. Some of the electronic gadgets for measuring velocity—now called chronographs—were something to behold, and they weren't really that much better than Robins' pendulum. An electric current did, however, offer the advantage of instantaneous signaling. In other words, with electric sensing devices activated by a passing bullet, there would be no error introduced by reaction time. Electric circuits were rigged so that the current was started when a bullet passed through a starting "gate." When the bullet "broke" or passed through a second gate, the current was instantly stopped. Now the only problem still to be solved was the same that

had plagued Newton: how to accurately measure the interval of time between passage through the two gates.

The principle of timing a bullet does not differ from the one we use to time the speed of a horse or a race car. We simply use a stopwatch to measure how long it takes the horse or car to speed over a measured course. When we know how far they traveled and how long it took, we can precisely determine their velocities. But when a bullet is traveling at supersonic speeds, a super stopwatch is needed to time the interval.

No such stopwatches existed even as recently as the first decades of this century, so other means had to be devised. One method was simply to compare a bullet's "gate-to-gate" time interval with an object traveling at a known velocity. Though I say the method was "simple," the procedure actually tended to be complicated, expensive, time-consuming, and, almost comical.

RUBE GOLDBERG DEVICES

Consider, for example, the "spinning chronograph," which utilized a paper disk made to spin at a known speed. When the bullet passed through the timing "gates," it caused a surge of electricity to arc through the spinning paper disk. By measuring the distance between the two holes burned by the electric sparks, the time lapse could be calculated with a reasonable degree of accuracy. As you would imagine, however, one of the problems was determining that the disk was revolving at a precisely known speed. If the revolutions were not exact, all the other calculations were in error, too.

The spacing between the two gates was usually 100 feet for slow bullets and 150 feet for the hotshot calibers. This generous distance between the gates tended to minimize errors in calculations, but at the same time, it presented a problem because the resulting velocity calculation revealed only the bullet's *average* velocity as it traveled from gate to gate.

The instrument that became more or less the industry standard until after World War II was the Le Boulengé. Though beautifully made with turned brass parts and wood cabinetry, it was as bizarre as anything you're liable to see in a Grade B science-fiction movie. The operational concept of the Boulengé was that gravity is a constant force and that a falling body represents a definite and known speed against which to compare the time lapse between points of a speeding bullet. And here's how it worked.

A metal rod more than two feet long was coated with soot. The rod was held in a vertical position by an electromagnet. When a bullet broke the first gate, which was a thin wire, it interrrupted the current to the electromagnet, causing the magnet to release the soot-coated bar.

When the bullet broke the second gate, it stopped current to a second electromagnet, which, in turn, released a weight that released a spring-powered knife blade. The knife blade swung, and it whacked the falling rod, making a distinct mark in the soot. The distance the rod had fallen before being whacked was then measured, and by using conversion tables, the bullet's speed was more or less determined.

When I was a teenager, I tempted fate on a regular basis by designing and building a blood-curdling assortment of really wild wildcat cartridges. Ever hear of the Carmichel Super Saber, circa 1952? I thought that my cartridges were launching bullets at somewhere near the velocity of light, but had no way of knowing for sure except by optimistically guesstimating their speeds. Like most wildcatters of that era of macho ballistics, I didn't own a chronograph, for the simple reason that they were very expensive. One dreary day, however, I came into possession of a surplus Boulengé chronograph, and I spent the next several weeks trying to make it indicate the velocities I was convinced my loads were generating.

It was a slow and tedious instrument. Though using one was an education in Rube Goldberg mechanics, it wasn't worth the effort. I finally traded it for something more useful—a set of used post-hole diggers.

The technological fallout of World War II and then the space programs totally revolutionized chronographs. Scientists had long been aware that when certain substances are activated by an electric current, they vibrate at an extremely fast and constant rate. Obviously, this was nature's own super stopwatch, exactly what was needed to measure tiny intervals of time. The problem, however, was devising a means of counting the vibrations. Once that could be accomplished, the remaining mysteries of bullet velocity would fall away like a maiden's final veil.

During the past few decades, devices have been perfected that count millions of impulses per second. With such instruments, it is possible to precisely measure the gate-to-gate time interval of a high-velocity bullet and also to radically shorten the spacing between the two gates. This allows simpler operation of chronographs, and it allows more precise recognition of a bullet's speed during a very short segment of its flight.

SOPHISTICATED MEASUREMENT

The earliest versions of these new velocity-measuring instruments were called counter chronographs. They were quite expensive and far beyond the means of most hobby handloaders and amateur experimenters. By the 1960s, the prices of compact, but excellent, units had dropped to a few hundred dollars. For the first time, it became possible for a handloader to conveniently measure the speed of his carefully concocted handloads or to check the actual velocity of factory-loaded ammunition fired in his own guns. One result of this was that wildcatting rapidly declined in popularity when it was discovered that most wildcat velocities were seldom as high as had been anticipated.

The early counter chronographs didn't offer a simple velocity readout like today's instruments. Instead, they were read by counting the illuminated lights on an instrument panel and inter-

---

CHRONOGRAPH MANUFACTURERS AND SUPPLIERS

Competition Electronics, Inc.
753 Candy Ln.
Rockford, IL 61111

Custom Chronograph Company
Rte. 1
Box 98
Brewster, WA 98812

Oehler Research, Inc.
Box 9135
Austin, TX 78766

Telepacific Electronics Company, Inc.
Box 1329
San Marcos, CA 92069

Tepeco
Box 342
Friendswood, TX 77546

Quartz Lok
13137 N. 21st Ln.
Phoenix, AZ 85029

preting the signals by means of conversion tables. This sounds rather complicated, and perhaps it was, but compared to the instruments of past ages, these chronographs were simple and speedy. With my first counter chronograph, I could determine the average velocity of five shots in 15 minutes or so. Most of this time was spent resetting the gates—a thin wire at the first gate and a printed circuit pattern at the second gate. Of course, if the bullet missed a gate, which occasionally happened, there was no reading. The reason for taking five or more readings for each load was that shot-to-shot velocities are seldom uniform, making it necessary to establish the average velocity of several shots.

The man who almost single-handedly transported velocity measurement into the space age was Dr. Ken Oehler. Though the word "wizard" is much overused, no other word will suffice in Oehler's case. He was a visionary who realized that chrongraphs could be small, easy to use, and inexpensive enough to appeal to even casual experimenters. The result of his engineering feat was the Model 10 Chronograph, which sold for an amazing $89.95 in 1967 and was guaranteed to be accurate to within an astounding 15 fps for bullets traveling as fast as 4,000 fps.

Initially, the shooting industry refused to take Oehler seriously. The notion that a chronograph could be sold for less than $100 was hard enough to swallow, but Oehler's accuracy claims had to be the irresponsible statements of a fly-by-night quick-buck artist. Surely, after bilking a few innocent shooters, he would disappear. But Oehler didn't disappear. In fact, he showed up on the doorsteps of ballistic laboratories and offered to compare his unit with the expensive and sophisticated chronographs used in commercial ammunition testing and research. More than a few skeptics discovered that the readings of Dr. Oehler's little chronograph matched those of their fancy laboratory units. When there were significant differences in readings, Oehler gently informed the doubters that *their* equipment was not performing accurately. More often than not, he was proved to be correct.

The one great advantage that expensive laboratory-type chronographs still had over portable hobby type units was the convenience of their photocells or "electric eye" gates. Hobby units such as Oehler's Model 10 required that the bullet physically penetrate and break the on-and-off gates. The photocell units were triggered when the eye "saw" the passing bullet. This was wonderfully convenient because it eliminated the tedious and time-consuming chore of setting up new screens before every shot. Of course, such laboratory equipment was mighty expensive. Even the cheapest models cost hundreds of dollars. Electric-eye choronographs would have been a blessing to handloaders and amateur ballisticians.

Oehler was again equal to the challenge. By the eary 1970s, he was marketing photo screens that were within the financial reach of anyone who could afford a rifle. But Oehler didn't stop there. Though he had broken the price barrier, Oehler's photo screens left something to be desired. They were rather bulky, as were all other such screens. Another drawback they shared with other photo screens was their reliance on a plug-in source of electricity. At many shooting ranges, there was (and still is) no electricity.

Oehler's solution created another revolution in the chronograph industry. In wizardly fashion, he reasoned that if Mother Nature supplied the necessary light, a photo screen could be truly portable and even cheaper than his existing photo screens. The result was the Sky Screen, which is now the industry standard for reliability and accuracy. No bigger than penny boxes of matches, these Sky Screens send their start and stop signals to the chronograph when they detect the *shadow* of a bullet speeding over them.

Today, Oehler's Sky Screens are connected to his fifth-generation chronograph—the Model 33. No bigger than a thin cigar box and utterly portable and battery operated, the Model 33 gives an instantaneous feet-per-second readout every time a bullet passes over the screens. At the touch of a button, it also tells the highest and lowest velocity readings of a string of shots. It then tells you the *average* velocity of all shots fired and the standard deviation of the velocity series (a helpful tool in evaluating load performance).

In 1950, such an instrument would have cost hundreds of thousands, if not millions, of dollars. Today, it costs less than $400. Though Ken Oehler stands as a symbol of chronograph development and quality, today's handloader can choose from several truly excellent and easily affordable chronograph units. Several makers are named in an accompanying list.

When considering the utility and convenience of, say, an Oehler Model 33 Chronograph, one wonders if the science of velocity measurement has gone as far as it can go. I think not. I believe that the genius of Robins and Oehler and others like them have only brought us to the threshold of ballistic knowledge. There are still voids in our understanding of a bullet's flight, and tomorrow's chronographs will strip away more of the mysteries. Already there is talk of laser, radar, and Doppler chronographs that could continuously record the speed of a bullet throughout its flight. Just think—you'll be able to press a button and see how fast your favorite load is traveling at, say, 160 yards or 230 yards or any other range!

Computer Programs for Shooters, Hunters and Reloaders

Howard E. French

As a writer, I took to the computer as a new-found friend—although, I must admit, in the beginning there were times when my family sometimes wondered if computer, screen, and printer might end up in the trash. Learning how to use the computer can be a bit touchy! Now, as a shooter and reloader, I have found the computer programs designed for shooters to be helpful tools I use more and more often. I have also discovered that nearly all the ballistic-type programs designed for the shooter are "user-friendly."

This means you don't have to be a computer whiz or memorize tons of commands to run these software programs. For the most part the screen tells you what to do and then shows you the results on the screen, or prints out the information for your records. You do have to have some information at hand, such as muzzle velocity, bullet weight, bullet coefficient, and other pertinent notes. In some cases the software program can supply you with some of this information, such as muzzle velocity for factory loads, and even coefficients of various bullets from many manufacturers.

Looking at five different software programs I have used, I am amazed at what they can do to help you and save a great deal of time and energy. If you are not familiar with computers, I should explain that the term "software" defines a floppy disk you insert into a computer (the computer is also called the "hardware"). When you purchase a ballistic program, the floppy disk has the complete program you purchased magnetically recorded on it. If anything happens to the software disk—your dog eats it, you pour coffee on it or bend it—the program is destroyed. If possible, make a copy of your disk, squirrel away the original, and use the copy. If you do destroy the original disk, most manufacturers will give you a copy for a moderate sum, provided you return the original disk to them.

USING SOFTWARE INFORMATION

Basically, once you have purchased the software, put it in your computer, turn it on, and you are ready to use the information encoded on it. The Sierra Ballistic program requires that on an Apple computer you first boot (enter into the computer) your Dos 3.3 System Master disk, before you enter their program. This is required only for Apple, not for the other makes of computers that can run the Sierra Ballistics program.

Once the program is shown on the screen, you can pick the particular part of the program you wish to use, and ask for it be implemented, This usually means entering a letter or number to bring the program into action. The following are some of the programs computer-designed to help you:

1) **Down-range velocity.** Determines the velocity of the bullet at various ranges you have set up for the chart. Normally you can set the differences in range, such as every 50 yards. Often you can use meters as well as yards, if desired.

2) **Flight time.** Tells how long it takes the bullet to reach each point on your table.

3) **Drop.** Determines how far the bullet drops below the muzzle of the gun if the barrel is parallel with the ground.

4) **Bullet path.** A useful piece of information as it determines the path of the bullet *when the sights are set to a certain range.* You enter the height of your sight, scope or iron, as well as the sighted-in range. This information will tell you the path of the bullet with this zero.

This article first appeared in American Shotgunner

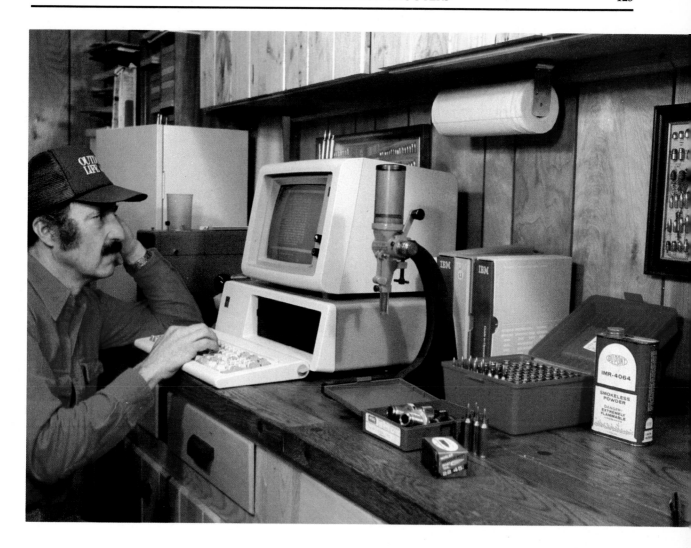

5) **Bullet-path comparisons.** Allows you to find out how loads of different velocities or different ballistic coefficients compare.

6) **Wind deflection.** Tells how far the bullet is moved by wind.

7) **Up/downhill shooting.** Indicates where you should aim your rifle for this type of shooting.

8) **Compute ballistic coefficient.** Coefficient is the factor relating the drag deceleration of a bullet to the drag deceleration of the standard bullet. The coefficient of your bullet is necessary to run some of the charts.

9) **Bullet energy.** Shows energy of the bullet at various ranges.

10) **Air density.** Allows you to determine where to aim at different heights. If your rifle was sighted in at sea level, you could determine where it would hit when fired at 6,000 feet in the mountains.

11) **Recoil energy.** Determines energy of various arms after you supply weight of gun, veloc-

Computer programs for shooters come from several manufacturers and can be used on many different home and office computers.

ity, weight of bullets, amount of powder, and other pertinent factors.

12) **Leading targets.** Provides guidelines on where to aim at moving targets.

13) **Compute reloading data.** Supplies correct powder charge for both factory and wildcat cartridges as well as expected chamber pressures.

ADVANTAGES OF COMPUTER PROGRAMS

Some readers may say much of this information is available in reloading manuals and from other sources. They are quite correct! However, you must look up this information, and it is usually somewhat swallowed in a sea of type. With the computer, you can enter your own information and have it printed out so it can be kept with your loading data. In addition, you can enter your own data from either factory loads you have chronographed or from wildcat loads.

In addition to those options listed, the computer programs can do other functions as well. One can convert inches into millimeters or vice versa; take figures listed as "¾" and turn these into decimal figures; let you turn inches on the target into clicks on your scope. In short, the more you use these programs, the more you find out how they can help you.

COMPUTER MARKSMAN

A unique program called The Trainer/Simulator is available from Computer Marksman. At first this looks a bit like a game; however, the more you use it, the more you find out how it will help your game shooting. First you pick the gun you will be using, and there are a number of different calibers listed, or you can enter your own load. You can then tell the computer whether you are using iron sights, or the degree of scope magnification. You can also specify whether you are shooting up or down hills and pick the maximum wind speed. You also enter

SOFTWARE FIRMS

Computer Programs	Computer Ballistics	Computer Shooter	Load From a Disk I	Pabsoft Ballistics	Sierra Ballistics
Down-range velocity	●	●	●	●	●
Flight time	●	●	●	●	●
Drop	●	●	●	●	●
Bullet path	●	●		●	●
Bullet-path comparisons	●				
Wind deflection	●	●	●	●	●
Up/downhill shooting	●				●
Compute ballistic coefficient	●		●		
Bullet energy	●	●	●	●	●
Air density	●	●	●	●	●
Recoil energy	●		●		
Leading targets	●				
Compute reloading data			●		

These programs are so complex that only a small part of their capabilities can be shown on a chart.

COMPUTER SOFTWARE AND TYPES OF COMPATIBLE HARDWARE

	Computer Ballistics	Computer Shooter	Load From a Disk I	Pabsoft Ballistics	Sierra Ballistics
IBM		●	●	●	●
Apple	●		●	●	●
TRS-80			●	●	●
Commodore			●	●	
Atari				●	
Vic-20				●	

Some of these programs require certain models, some work on IBM compatibles; check with manufacturer.

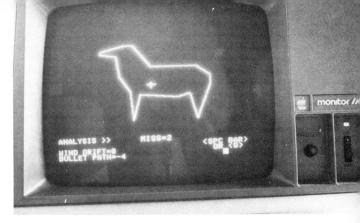

Trainer/Simulator from Computer Ballistics (left) allows you to try your marksmanship at various ranges. Target here is deer at 250 yards as seen through 4X scope with rifle sighted-in at 200 yards. After you fire your shot with Trainer/Simulator, target enlarges (right) so you can see where you hit. Unfortunately, this shot was slightly low.

your zero range and the maximum range. To mix things up, the computer will show different ranges, different wind speeds and directions, and also different amounts of hill angle. You can also pick the animals you are hunting, from varmints to deer, elk, or bear.

The computer will then show you the target, varying in size due to the range or the power of your scope. The direction and speed of the wind is noted and an uphill or downhill angle is presented. You can adjust the aiming point by using "L" for left, "U" for up, etc., and then "fire." The animal is then enlarged to full-screen size on the monitor, and your "hit" is shown. If you missed, this is also shown, and you know if you hit too high, too low, or misjudged the wind. It is a valuable tool and a lot of fun as well.

LOAD FROM A DISK

Load from a Disk is a very useful program that can help with loads for factory or even wildcat cartridges. I tried it out on the 7mm Bench Rest round, and was surprised at the results. Load from a Disk came up with one load just .2 ($^2/_{10}$) different from a load I had been using, and the velocity difference was only 67 fps different from my chronographed load! Naturally, this program is for the careful shooter: you are warned to look up data in a loading manual and start using the lowest load furnished.

In addition, W.W. Blackwell, the developer, also has another program called Load from a Disk II. This has five different sub-programs: Optimum Rifle Twist, Muzzle Velocity from Trajectory, Ballistic Coefficient from Velocity, Ballistic Coefficient from Trajectory, and Ballistic Coefficient from Bullet Shape.

IN CONCLUSION

I am afraid this article would become too laborious if I were to detail the extra little bits of information these various programs have to offer. But if you have a computer or are thinking about getting one, keep these programs in mind. They will be of immense help to any shooter, hunter, reloader, or serious-minded experimenter.

The "Inside" Story . . .
Big Game Bullets

Ross Seyfried

When you read this, you will see several references to Africa, and many photos of large-caliber bullets. I don't want anyone to think I have lost sight of our American hunting and the great game animals of home. Africa is my learning place. While I don't shoot much myself, I get to see several lifetimes of hunting each season, and what I learn there about bullets is too valuable not to pass along. Many of the photos are of large-bore bullets simply because it is easier to see the construction of a big bullet in a photo. Most of the manufacturers make a wide selection of small-bore bullets too. My reason for writing this is to give everyone a better chance of being successful when they take the shot of a lifetime. This might be at the biggest trophy whitetail ever, or your son's first buck.

The first big-game bullets were chipped flint, tied to the end of a spear or arrow shaft. Progress took a giant step forward when men learned to load lead balls into rifles and launch them with charges of black powder. I can conjure up visions of the greatest hunter, Selous, hunched over a fire casting quarter-pound balls of lead, getting ready to face elephants the next day. He used the latest bullet technology, adding mercury to the lead, producing hardened bullets. This gave him better penetration and a greater margin of safety when he took on the great game animals that he hunted. What is a wonder is that the mercury didn't kill him, cheating the elephants out of any chance. Selous did live long enough to see the dawn of modern high-velocity rifles and their jacketed bullets. When Selous was carrying his huge 4-bore elephant rifle, a .577 was considered a *light* deer rifle. But with the invention of jacketed bullets, the .577 became a better elephant stopper than the giant round-ball guns. The modern rifle saw a little refinement over the next 100 years, and a raft of new cartridges was invented.

But the jacketed bullet really hadn't changed too much from the first ones that Selous fired before the turn of the century.

With a few notable exceptions, bullets were lead cores covered with a copper alloy jacket. Some bullets were open on the front end (soft noses) and some were open on the back end (solids). Some soft-nose bullets performed well, some blew up, and some didn't open at all. Similarly, some solids did their job well and some didn't. When they didn't, sometimes a hunter lost his life.

Very little was done to change the reliability of jacketed bullets until John Nosler offered his Partition bullets in 1948. From there things stood relatively still until about three years ago when hunters began to seriously ask for bullets that would perform better. Bullet makers large and small began to turn out a crop of the finest hunting bullets ever produced. Both softs and solids are getting attention.

A "solid" is a bullet that is designed to absolutely hold its original shape when it hits the game. A good solid won't dent, bend or even scratch badly even though it often crashes into the heavy bones of elephant, buffalo, or hippo at point-blank range. When a hunter is taking on big, dangerous game, there are only two kinds of solids—the very best and all others. I don't want anything to do with the "all other" category. They can get you killed!

In defense of our American bullet makers, they have been supplying solids with steel jackets that do a very good job. Unfortunately, the steel-jacketed solids will fail. Where there is no room for failure, three bullet makers have set to work to produce "flawless" solids.

Barnes, Trophy Bonded Bullets, and A-Square

This article first appeared in Guns & Ammo

Top two rows of bullets are shown in cut-away views to reveal various approaches to design and construction for use on various types of game. In bottom row are bullets recovered from game.

Labels for top two rows and bottom row:

Winchester Silvertip .338—250-grain

Trophy Bonded "Bear Claw" .416—400-grain

Hornady FMJ .458—500-grain

Barnes Super Solid .416—400-grain

Barnes Soft Point .458—500-grain

Nosler Partition .338—250-grain

Trophy Bonded "Sledgehammer" .458—500-grain

DWM H-Mantel 8mm—198-grain

Swift .375—300-grain

Sierra HP Boattail .30—180-grain

Trophy Bonded "Bear Claw" .375—300-grain

Sierra .375 300-grain SPBT elk/250 yards

Hornady .458 500-grain SP Cape buffalo/60 yards

Federal Premium .30 150-grain SP deer/150 yards

Nosler .338 250-grain Partition eland/550 yards

offer a range of solids that are virtual perfection. The Barnes and the A-Square solids are made of homogeneous material. They are turned on a lathe out of solid bars of a bronze-like alloy. I have used the Barnes "super solids" extensively. At first I was using them in my .416 Rigby. With the big Rigby case and its 2,500-fps-plus velocity, I was never able to recover a bullet out of a Cape buffalo. They simply went right through the buffs from any angle and, I might add, killed them one and all. This year I used a rifle chambered for the .416 Taylor cartridge. Its velocity was sufficiently lower than the Rigby that some of my solids stuck under the hide on the "far side" of some buffalo. I also used the A-Square Monolithic solids this year and recovered some

of them too. The results were the same—bullets engraved by rifling, but otherwise unmarked.

The A-Square bullets are available in the small bores as well as the heavy calibers. I was able to watch .30-caliber 180-grain solids hammer two buffalo this year. I took one lad of 12 after buffalo and let him use my .300 Winchester loaded with the 180-grain A-Square bullets. Randolf shot his bull in the center of the shoulders. The little bullet flattened the buff, which got up and ran 50 yards and died. I didn't get to look at the bullet; it went through the bull like he was made of Swiss cheese and into the mopane trees beyond. *G&A*'s Women's Shooting Editor, Jo Anne Hall, shot another buffalo with the same combination. Her bull was facing her when the 180-grain solid hit the point of his shoulder. This bull went to his knees, then ran some 100 yards before he expired. I found that bullet under the skin on the buffalo's hip! You can see the pictures of this bullet. It could easily be reloaded and fired again. I have seen these "homogeneous" bullets of Barnes and A-Square used in calibers from .30 to .500 and the results are always the same. Unlike other conventional solids, they don't bend, flatten, or occasionally blow up.

Because they are made of a material that is less dense than lead, these bullets are longer for any given weight than a solid with a lead core. There are times when this extra bullet length can cause problems. The .458 Winchester, with its all-too-small case capacity, is one of them. Barnes makes a lighter .458 bullet that works well, and there are the Trophy Bonded solids that still use a lead core and are the same length as conventional bullets. The Trophy Bonded solids still use a lead core, but that is where the similarity ends between them and conventional solid bullets. The jackets are super-thick naval bronze, and the core is dovetailed into the bullet's base. From what I have seen, these bullets are virtually as indestructible as the homogeneous bullets. They also have the advantage of conventional length and a flat nose. This flat nose should lend itself to adding punch and to keeping the bullet pointed straight forward, just like the Keith shape does in handgun bullets.

Of course, everyone wonders why I care so much about solids, with their limited uses. They actually are a lot more useful than you might think. While solids are primarily designed for use on big, dangerous game, they are extremely useful when you hunt any kind of big-game animal. I almost always carry a few solids for any rifle I hunt with. There is nothing better for dispatching a wounded animal than a solid. If you are unfortunate enough to have wounded either a deer or an elk and have to follow it, chances are that the shot you will get to finish your quarry will be going away. A good solid planted on the seat of the pants will drop the animal almost instantly by penetrating completely through, lengthwise! If you get a broadside shot at a previously wounded animal, you will be surprised at how quickly a solid bullet works when properly placed on the shoulders or ribs. The other bonus is that a solid doesn't make hamburger out of good steaks.

Whatever your particular application might be, you now have solids that can be absolutely trusted to do their job. I would like to get the ammunition companies that load solids to look at these bullets. Their rounds are loaded primarily for use against dangerous game, and the bullets they are loading now fail miserably on oc-

Importance of big-game bullet performance is most critical when hunting dangerous game. Even though his quarry is down, this hunter approaches buffalo with extreme caution.

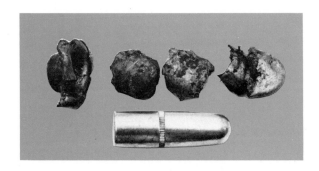

Unfired 300-grain .375 factory solid is pictured with "remains" of the same make after it blew up on shoulder of buffalo, having penetrated only 6 inches.

Two bullets at left are Barnes .416 Super Solids, fired and unfired. Fired specimen was recovered from buffalo. Also shown are unfired .308 180-grain A-Square and similar bullet fired by Jo Anne Hall through Cape Buffalo, penetrating from shoulder to hip.

At moments like this, bullet failure could result in extreme disappointment if not outright disaster.

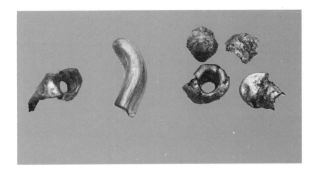

These three "conventional" solids all failed. From left: exploded steel jacket; bent copper-jacketed solid; and all that remained of factory .375 solid.

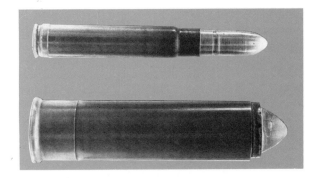

Looks can be deceiving. Relatively small .416 Taylor loaded with Barnes Super Solid may be more effective than old 8-bore.

casion. If they fail when a man's life is at stake, the price is far too high. You can see just what I mean from the photos of solids that have failed. I collected all of these bullets over the last two seasons, so the frequency of failure isn't all that low. At least one of these failures could very easily have led to one of my trackers' demise by a buffalo. The ammunition companies have always loaded the best solids available until recently. It may be time for a change.

If the bullet maker's skill with solids has increased a little, the advances in soft-nose bullets should be measured in leaps and bounds. I categorize soft-nose bullets in four different layers: the finest handmade bullets, Noslers, commercial bullets that have been designed to retain their cores and perform well on big game, and "the rest of the bullets."

In any conversation about great handmade bullets, the Bitterroot bullets have to be mentioned. They have been around a long time and in a variety of calibers. They are the first bullet that I know of to employ a physical bonding between the lead core and the jacket. The Bitterroot bullets are very tough, and at times have difficulty opening or expanding at low velocities. But they won't shed their cores or blow up. Unfortunately, the Bitterrots are in very short supply and almost impossible to get.

The new Trophy Bonded bullets are similar in some ways. They employ a jacket turned out of copper on a lathe and the core is physically welded into the jacket. The Trophy Bonded bullets are probably the most vicious bullets I have ever seen. When they expand they look like fan blades, and the damage they do is spectacular. The bullets will expand on a whitetail's ribs and yet will shoot right through most of a Cape buffalo on a broadside shot. They will almost always retain at least 90 percent of their original weight. The Trophy Bonded bullets are available (yes, I mean they are available) in sizes from the 7mm's right through the big boys, with more sizes being added all the time.

The next great bullets are those made by the Swift Bullet Company. If I had to pick a best big-bore soft-nose bullet, this is it. The Swift H-Frame Safari bullets may be the finest soft noses ever made. The jackets are pure copper with an H-Frame that divides a front and back core. The front core is bonded to the jacket. These highly sophisticated bullets begin to expand on even the thinnest skin and roll back to a huge, classic mushroom shape. It's difficult to believe, but I would gladly use the Swift .375 or .416 bullet for a broadside shot on a leopard or a Cape buffalo. That says a lot. A leopard is one of the softest animals anyone will ever shoot, but the bullet has to expand immediately and violently to give the desired results. The buffalo, on the other hand, is one of the "hardest." Getting a soft nose through the vitals on old Nyati is a real chore. I have seen the Swift .416 smash one shoulder, chop a 3-inch hole in the spine, and pound the off-side shoulder before it quit—yet retain 380 of its 400 original grains in the process. The Swift bullets are made in .375, .358, .416, and .458 only. If the Swift and Trophy Bonded bullets have a drawback, it is that they expand to such huge diameters that maximum penetration may be a little less than desired. But for every normal critter alive and broadside shots on buffalo, they are absolute tops.

There is one more entry into the superstar class of soft-noses. This is a brand-new Barnes bullet that is taking a different approach to the penetration and weight-retention problem. This bullet will be almost all *jacket*, with a tiny lead core in the nose that is considered "expendable." These bullets will be like solids with noses that will expand. I am very excited about the new Barnes bullets. In practice they may prove to be the best of both soft and solid. Some of the best news is that they aren't just for big-bore fans and handloaders. The new Barnes softs will be loaded in PMC ammunition in calibers from .243 to .300 Winchester Magnum. Watch for them; they may be the finest soft-noses ever made.

When I talk about bullets, I classify the Nosler *Partition* bullet by itself, simply because the performance deserves such special mention. I have never had a Nosler Partition bullet fail me. The partition dividing the two cores lets the first core

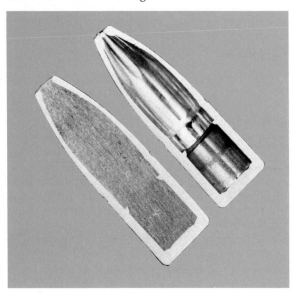

Speer uses novel approach with its Grand Slam line of bullets (shown sectioned) involving dual core hardness and internal rear-core ring.

expand immediately and violently back to the partition. The front portion of a Nosler is so fragile that it will still expand reliably at very low velocities or on soft lung tissue, while its rear portion drives forward like a flat-nose solid. I hear the complaint that a Nosler sheds too much of its initial weight. I admit that they lose a lot of weight when the front section blows off, but they do a lot of damage and penetrate as deeply as any soft-nose I have ever fired. What Noslers *don't* do is fail to expand or blow to bits, and that's exactly what a bullet is supposed to do. Another plus is that you can get the Nosler Partition bullet in a few selected Federal Premium loads. This is the best "over-the-counter" ammunition available to date.

This year I watched one of my clients put on one of the finest game shooting exhibitions I have ever seen. He came to Africa badly undergunned in my opinion. He carried a very old and worn Winchester Model 70 .270 and Federal Premium 150-grain Nosler ammunition. Bill was a fantastic game shot, and I am sure he knew this beloved old .270 as well as he knew his wife. He and his grandsons took animals ranging from lion on down, almost every one cleanly with a single shot. The lesson: Not even horsepower is equal to bullet placement and *bullet performance.*

The next category of soft-nose bullets is the good, mass-produced game bullets. The Winchester Silvertip and the Remington Core-Lokt are the old standby game bullets. They haven't changed appreciably for years, but both do well in keeping the cores in the jackets, expand reliably, and generally do a fine job on game. The Hornady interlock bullets are a fine example of a manufacturer really caring about making a superior bullet. When Hornady added a little internal ring in the jacket to all but eliminate any chance of core separation, it took what I considered a very ho-hum game bullet and made it a great performer. This little change cured what I think is the most common fault of jacketed soft-nose bullets—that is, the core coming out of the jacket to be ground to bits in the harsh world of high speed and big bones. These fine Interlock bullets are available for the handloader and also in ammunition form in Frontier ammunition.

The Speer Grand Slam bullet is another example of a response by a manufacturer to our request for bullets designed to take big game. The Grand Slams are very sophisticated bullets, with two different hardnesses of lead in the core. The front part is very soft, and the rear part of the core is hardened lead, held in place by an internal jacket ring. Again, the emphasis is on a bullet that will expand, but not fly apart. I have had very good success with the Grand Slams on game. They aren't the same as a handmade bul-

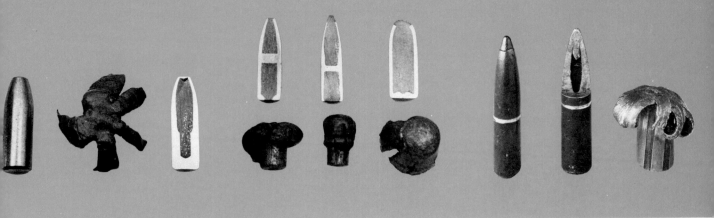

Examples of superb bullet performance! In photo at left is Trophy Bonded 500-grain .458—complete, fired into buffalo, and sectioned. Center photo shows sectioned and recovered 400-grain Swift .416, 250-grain Nosler .338, and 400-grain Barnes .416. In photo at right, you see new Barnes X-HC copper bullets. Fired specimen is 270-grain .375 recovered from brown bear. Others are 180-grain .308s, complete and sectioned. All of these photos illustrate performance hunter needs and expects from premium bullets.

let, but they are head and shoulders above any ordinary bullet not designed specifically for taking game animals.

The last entry in our good bullets is the ordinary, plain vanilla Barnes bullets. They are pure copper tubing jackets with pure lead cores. As ordinary as they appear, their performance is not at all ordinary. These are very good game bullets. The pure soft copper and lead tend to expand very reliably, and their "mushy consistency" tends to make them ball up and not fly apart. The Barnes bullets are available in almost every conceivable caliber up to .600 Nitro.

It isn't possible to cover every bullet individually, but most of the rest of them should be considered also-rans. They are the bullets with conventional jackets and cores that may or may not perform the way we need them to in order to take big game cleanly. Boattail bullets should always be viewed with high suspicion. I think the world is suffering from boattail mania. I realize that they give better ballistic performance at long ranges, but the tapered core inside the boattail jacket is an open invitation to core separation, underpenetration, and wounded animals. There is nothing wrong with the boattail concept; it's just that boattails should have some special means of holding the core in the jacket, like the Hornady Interlock, if they are going to work.

One kind of bullet that shouldn't be used on game is the hollow-point match bullet. These are bullets specifically designed for superior accuracy on targets. They are not intended for shooting game animals. They may or may not expand, and if they expand, they may or may not hold together—far too many "may or may nots" when we are talking about the possibility of wounding

an animal. Don't let that golden gleam of accuracy and long-range ballistics cloud your clear view of the necessity of bullet performance where it counts . . . after it hits meat.

There are two factors that shouldn't carry too much weight when you select your bullets for hunting: cost and gilt-edge accuracy. When you consider how much a game bullet costs, first consider what a tiny portion of the cost of *any* hunt will be spent on bullets. If you use the most expensive bullets in the world on a full-blown safari, chances are you won't have used $50 worth when the trip is over. On the average deer or elk hunt, $5 worth of bullets that cost $1 each will have been a hell of a lot of shooting. You can use any kind of bullet for your practice and initial sighting in, then switch over to your game bullets to check your zero and for the actual hunt.

How accurate a bullet needs to be is an interesting question. I am as guilty as the next guy of wanting all the accuracy I can get. Almost every bullet mentioned here will stay under 2 inches at 100 yards. Most will group a lot better than that. My advice if you're going hunting is to take a game bullet of known performance and 2-inch grouping every time over a "½-inch" bullet that might fail when it lands.

Exactly what game bullets you use depends on the kinds of animals you'll seek and the conditions under which you will hunt them. If you're hunting dangerous game with solids, there is no choice—use only the very best you can buy. With soft-nose bullets, the tougher and the closer your game is, the better your bullets will need to be. If there is doubt, use a better bullet than you think you might need. Your success and the game animals are worth the small difference in price.

PART SIX

WORKS OF ART

The Art of American Arms

R. L. Wilson and Richard Alan Dow

Some of the most beautiful objects of decorative art have come to us through the custom of embellishing functional items. Through applied art, the practical and serviceable are elevated to the beautiful and exotic. Nowhere has the tradition been more highly developed, more rooted in history, than in the engraving and enriching of fine firearms. The loving craftsmanship that gunsmiths have brought to handmade arms over the centuries invites equally loving and careful decoration.

Runic inscriptions on sword blades were believed to imbue their bearers with mystical power. Etched surfaces of graceful armor gave the appearance of rich fabrics of civilian dress. And the inlaying and engraving of handguns and sporting arms down to the present (an increasingly difficult art as metals have become harder and more resistant to the engraver's tools) lend an elegance that transcends their technical nature.

The methods of engraving firearms, and the tools used, have remained virtually unchanged since the seventeenth century. Slender chisels in various shapes, called scribers, are tapped with small jewelers' hammers to form the basic design and shading. The graver, a shortened scriber with a bulbous wooden handle, is used to produce fine detail. Vices hold the piece being worked, magnifying glasses permit the artist to see the minute detail, and drafsmen's dividers measure the correct dimensions. The rest depends on the skill and artistry of the engraver.

Gun engraving has spawned a host of masters. In Europe, the earliest makers employed engravers to embellish their handmade arms. The Italian firm of Beretta, the oldest industrial company in Europe and still in the hands of the founding family, has offered embellished arms since the Renaissance. The British firm of Holland & Holland, creator of coveted shotguns, dates from 1835.

In America engraving was standard on certain arms, especially the Long Rifle (the fabled "Kentucky Rifle"), and became a by-product of the machine age. Henry Deringer's pocket pistols, a

This article first appeared in American West magazine

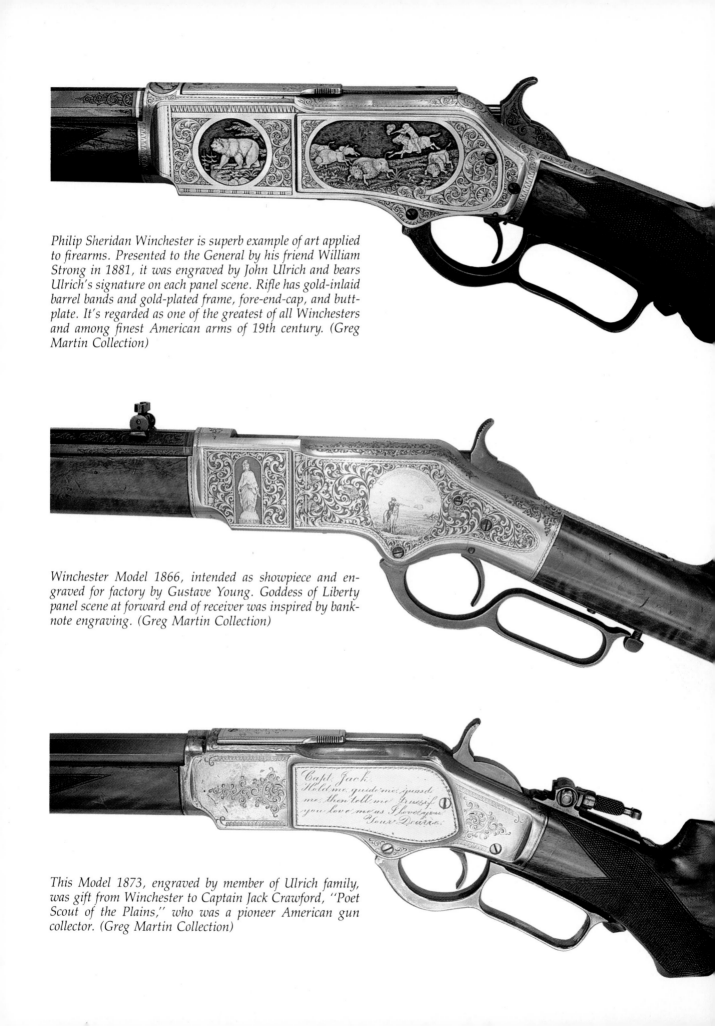

Philip Sheridan Winchester is superb example of art applied to firearms. Presented to the General by his friend William Strong in 1881, it was engraved by John Ulrich and bears Ulrich's signature on each panel scene. Rifle has gold-inlaid barrel bands and gold-plated frame, fore-end-cap, and butt-plate. It's regarded as one of the greatest of all Winchesters and among finest American arms of 19th century. (Greg Martin Collection)

Winchester Model 1866, intended as showpiece and engraved for factory by Gustave Young. Goddess of Liberty panel scene at forward end of receiver was inspired by bank-note engraving. (Greg Martin Collection)

This Model 1873, engraved by member of Ulrich family, was gift from Winchester to Captain Jack Crawford, "Poet Scout of the Plains," who was a pioneer American gun collector. (Greg Martin Collection)

Modern masterpiece, combining arts of gunsmith, engraver, stockmaker, and case builder, Safari Club International 1986 "Leopard" set was made by David Miller Co., Tucson. Gunmakers were David Miller, Curt Crum, and Dale Drew; engraving by Lynton McKenzie. Action is pre-1964 Model 70 Winchester. Elegant case was fashioned of South American ironwood and lined in gray ultrasuede with brown trim to hold rifle and its accessories in fitted recesses. Accessories were engraved by Steve Lindsay, while Ray Wielgus created gold inlay as well as antiquing of case's brass hardware. Knife was made by Steve Hoel. Leopard scene on rifle's floorplate was taken from especially commissioned painting by Guy Coheleach. Unique in its method of attachment, steel plate has no screws to interrupt engraving, and its fastening is not visible. Close-up reveals exquisite metalwork on action, receiver, bolt handle, and bolt knob. Steel buttplate, decorated with contemporary design unique to David Miller Co., commemorates 50th anniversary of Winchester Model 70. (Joe Bishop Collection)

Photography credits:

Sid Latham/Bruce Pendleton/G. Allan Brown/
Winston G. Churchill/Ronald D. Dehn/Linda Fry Poverman

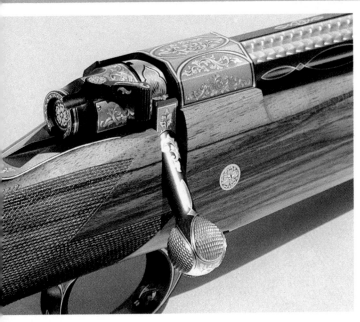

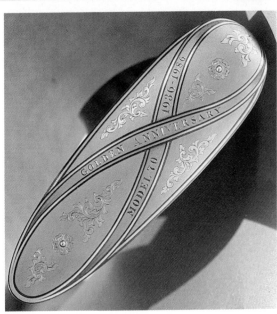

Here is another masterpiece of 19th-century gunmaking artistry by Gustave Young: pair of presentation 1860 Army Colt revolvers used as diplomatic gift by President Abraham Lincoln to King Charles XV of Sweden in 1863. A keen arms collector, Swedish monarch fired revolvers once, had them cleaned, and thereafter admired them as part of one of Europe's most magnificent arms collections. (The Royal Armoury, Stockholm)

product of hand craftsmanship and early machinery, were nearly always engraved. Mass-produced "pepperbox" multi-barrelled pistols of the 1830s and 1840s often sported hand engraving as standard, and the demand for specialist gun engravers was strong. When Colonel Sam Colt's first revolvers issued from his short-lived Paterson, New Jersey, plant, most of them were plain except for engraving on the cylinders, which Colt cleverly executed with roll dies, Only a few Paterson Colts were hand engraved, but they bolstered the demand for craftsmen able to execute such work.

Although most manufacturers no longer offered engraving as a standard feature by about 1850, deluxe-grade guns were maintained in inventory and produced to the special order of military and civilian clients. Samuel Colt was in the forefront of this development, promoting his business and seeking government contracts both in America and abroad through carefully calculated presentations of engraved revolvers to heads of state and to military and naval officers. Deluxe guns were made in Colt's new Hartford factory almost as soon as the doors opened in 1847. Sharps, Volcanic, and Smith & Wesson followed Colt's lead within a few years.

The American arms industry developed rapidly as a result of demands made by western migration, the Mexican War, and continuous battles with Indians, as well as events in Europe, South Africa, and Australia. At the same time, the Victorian taste for often garish and profuse decoration in art and architecture profoundly influenced weapons embellishment. A golden era of arms decoration quickly evolved; more than 250 gun engravers were active in the United States alone between 1850 and 1914.

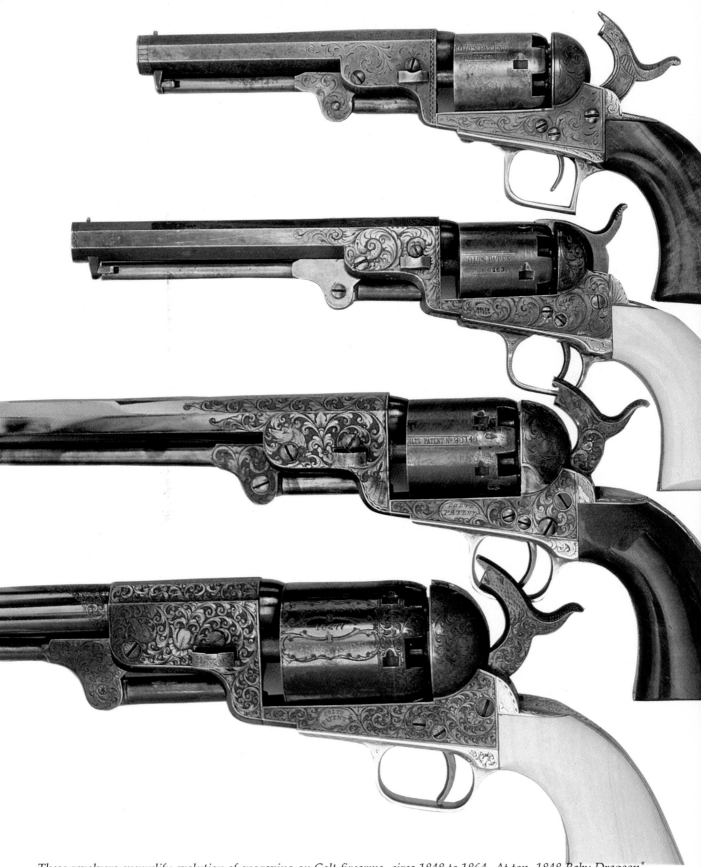

These revolvers exemplify evolution of engraving on Colt firearms, circa 1848 to 1864. At top, 1848 Baby Dragoon is adorned with ''early vine style'' scroll. Next, 1849 Pocket Model displays ''donut scroll.'' Third from top is 1851 Navy with Gustave Young scroll (introduced at Colt's in 1852). Finally, Third Model Dragoon displays Young scroll of the sort known as ''extra engraved.'' (Greg Martin Collection)

Hundreds of Germans and Austrians responded to the American demand for expert gunmakers, most of them settling in New England. The emigrations began in force in the 1840s and 1850s and continued in diminishing quantities into the 20th century. Several master gun engravers also came to America in the 1840s, among them Louis D. Nimschke, Gustave Young, and the first of the Ulrichs—men who became the most influential style setters and taste makers in the evolution of arms engraving. The Young and Ulrich families became dynasties, with descendants active in the craft well into the 20th century.

The rich, strongly Germanic style of these engravers still dominates American arms engraving and is popularly called "the American style." Among the better-known examples of this work are a Colt Single Action revolver done for Theodore Roosevelt ("his best Western revolver"), a Model 1876 Winchester presented to General Philip H. Sheridan, gold-inlaid Colt percussion revolvers presented to Czar Nicholas I of Russia, equally lavish Colts presented to the Sultan of Turkey and the kings of Sweden and Denmark, and a Remington Rolling Block rifle and Army Model revolver presented to General George Armstrong Custer.

The beginning of the 20th century saw a decline in the commercial arms trade. Most of the "Wild West" had been tamed, and the great era of big-game hunting was suspended until conservation and game management later restored many species in quantity. World War I interrupted civilian arms production, and, afterward, a shortage of craftsmen plus the ravages of the Depression saw the engraving of arms almost die out. Of equal significance were the advances in metallurgy, producing harder steels, many of them so hard that engraving tools often broke and lines were difficult to cut neatly or with depth. As a consequence, the time required to complete an engraving assignment increased.

Prominent during this period were R. J. Kornbrath, George Ulrich of Winchester, Wilbur Glahn of Colt, and Harry Jarvis of Smith & Wesson. Probably not more than 20 persons in America made their living from gun engraving between 1914 and 1945. The most influential was Kornbrath, an Austrian whose reputation has become almost legendary. In general, these engravers held much more closely to the German style than their predecessors had, and American engraved arms from this period differ little from their European counterparts.

A renaissance of interest in engraving began after World War II, which has resulted in a body of craftsmen unequalled in any field—even though one would assume that the best craftsmen would be in the jewelry trade. Many of this new breed were veterans returning from wartime service and looking for intriguing and challenging employment. A renewed interest in shooting sports, a generally favorable economy, and the realization among buyers that quality decorative firearms represent good investments have been strong influences. Furthermore, like the majority of arms collectors, most engravers recognize the field of arms and armor as unique in art and antiques in its extraordinary multi-faceted combination of art and craftsmanship, history, mechanics, and—especially in western Americana—romance.

The growing number of engravers at work in the United States is a reflection of America's fascination with its firearms heritage. Perhaps 200 Americans now make their living at gun embellishment, and who knows how many hundreds more work as part-time artisans. This new breed employs a wider than ever range of pictorial and scroll themes, heavily influenced by the centuries of gunmaking as well as the ability to share knowledge and technique with each other. A veritable library of books on arms engraving has appeared in the past 40 years, and decorated firearms are frequently featured on the covers of gun and shooting periodicals.

While not all those working today are masters, masters do exist, on both sides of the Atlantic. In the opinion of the authors, the finest work on firearms stocking, engraving, metalwork, and casing is being done *today*. Among the contemporary giants of gun engraving are K. C. Hunt and the Brown Brothers of England; Hans Obiltschnig of Austria; Wolfgang Tschinkowitz of Germany; Firmo Fracassi, Angelo Galeazzi, Gianfranco Lancini, Francesco Medici, and Gianfranco Pedersoli, all from Gardone Val Trompia Italy; J. Baerten, René Delcour, and his protégé Philippe Grifnee of Belgium; and Mateo e Cayetano and José Careaga of Spain. In the United States are A. A. White, a master of chiselled steel and gold inlay, widely recognized as "dean of American arms engravers;" Lynton S. M. McKenzie, an Australian who came to America via the London arms trade; Frank Hendricks of San Antonio; Leonard Francolini of Connecticut; Sam Alfano of New Orleans; and Winston G. Churchill of Vermont, who was taught engraving by Joseph Fugger of Abercrombie & Fitch's Griffin & Howe, New York, who in turn was a protégé of R. J. Kornbrath.

Alvin A. White, trained in the jewelry industry of his native Attleboro, Massachusetts, and later at the Rhode Island School of Design, was one of the first to respond to the post-World War II demand for high-quality decorated arms. A master not only in steel but in fine metals, ivory,

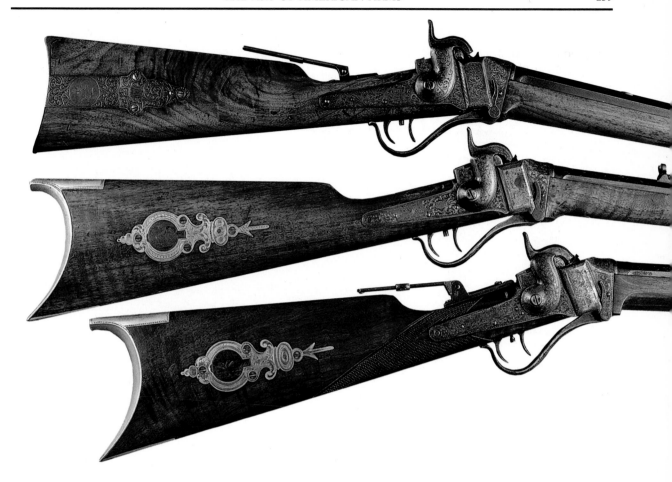

Produced in Philadelphia, Hartford, and Bridgeport, guns made by Sharps Rifle Company reflect their age of manufacture in quality and decoration. Most were plain service arms. Relatively few were deluxe pieces, boasting various degrees of engraving and stock work. Gustave Young, who so strongly influenced Colt engraving—from 1852 to the present—also decorated enough Sharps rifles (and guns by other makers) to dominate that firm's style of embellishment. These three Sharps rifles all show work of Young himself or his shop. (Greg Martin Collection)

pearl, and wood, White turns his creative genius to scrimshaw, knife making, silver and gold smithing, and inlaid cabinetry as well as gunmaking, stock carving, and engraving.

One of the most important developments in gun engraving occurred in the 1970s, when the Colt Firearms company opened its Custom Gun Shop and concurrently returned black-powder (muzzleloading) arms to its product line. As a result, some of the finest Colts ever made date from the past 15 years. Although the Colt black-powder series was discontinued in 1982, the supplier of most of those guns, Aldo Uberti of Gardone Val Trompia, Italy, continues his operation with the gusto of his 19th century predecessor, Colonel Samuel Colt. At this writing the finest Colt-style revolver ever built is in the works at Uberti's factory; it is a replica of the Walker model of 1847, gold inlaid to outdo the presentations from the Colonel to Czar Nicholas I in 1854.

The Colt Custom Gun Shop has created mod-

ern masterworks now in the collections of kings and presidents and the most discriminating of gun owners. Drawing on 150 years of tradition, Colt today offers its entire line of longarms and handguns in custom embellished form. To promote the art of engraving, it regularly creates masterpieces for special occasions and shows; many of these pieces are later sold at auction. Another American manufacturer following in the great tradition of deluxe arms is Sturm, Ruger & Co. The founder and chairman, William B. Ruger, has commissioned a series of deluxe guns of his own make, richly engraved and inlaid by his in-house artist-engraver, Paulus Lantuch. This growing collection, never displayed publicly, represents a body of work that is a salute to the history of arms embellishment.

Firearms have held a unique fascination throughout our history. Richly engraved arms have been, and will continue to be, the most sought after.

The Magnificent Obsession

David A. Webb

THE annual Shooting, Hunting and Outdoor Trade (SHOT) Show has something for everyone in the displays of hundreds of exhibitors. Costly and exotic firearms from around the world are there to be examined.

But a single stand caused more excitement than any other at the show's 1986 edition in Atlanta— a collection of 35 rifles stocked in strange and marvelous woods from every corner of the globe.

The "Woods of the World" display was a prominent part of the Reinhart Fajen booth. It was the brainchild of Dr. Joseph Iannicelli, a Georgia inventor/enterpreneur who has scoured the world to amass his unique collection of stock woods. Five illuminated cases held the .300 Winchester Magnum Browning BBRs, with stocks ranging from snowy bird's-eye maple to pitch-black ebony.

The genial Iannicelli displayed his collection and answered questions from hundreds of show visitors, including notables like pro football great and NRA Life Member John Riggins and former President Jimmy Carter. Carter was the only observer who got to handle an Iannicelli rifle without donning a pair of protective white gloves.

Iannicelli, an NRA Life Member, conceived the idea of a collection of different wood species more than a decade ago. As he tells it, he has always been fascinated by wood, its variations in color, and the figure and grain characteristics. His father had been trained as a cabinet maker and kept a fully equipped shop even after starting his own construction business. Apparently, his appreciation for fine wood and woodworking was passed on to his son.

The collection remained just an idea until Iannicelli teamed up with stockmaking legend Reinhart Fajen. The two decided that the work would begin with woods that Fajen had in stock. Joe Badali of Browning helped locate the first barreled actions, and the project got started with three maple stocks.

Iannicelli selected Fajen's Aristocrat model, a swoopy design with Monte Carlo comb, rollover cheekpiece, and contrasting grip cap.

The first three guns were the only ones in the collection that were checkered. Iannicelli saw that the checkering covered the wood grain, so the remaining rifles were left unadorned.

About ⅔ of the 35 woods in the display were procured by Fajen through his network of specialty wood suppliers. Woods like cocobolo and zebrawood were originally acquired for use as fore-end tips and grip caps for more prosaic stocks.

For his part, Iannicelli has made it a practice to track down exotic woods like ipe, sucupira, goncalo alves, padauk, and kingwood during his worldwide business travels. The quest is made more difficult by the fact that the Aristocrat design's rollover cheekpiece requires a blank 2½ inches thick. The standard lumber thickness throughout the world is 2 inches.

So Iannicelli finds that he is looking for rare woods in a nonstandard size. He always carries his current want list of woods, a tape measure, and an illustrated handbook on woods to bridge language gaps. The common names of woods not only differ in various nations, but the English name often means an entirely different wood abroad. When the going gets tough, matters are resolved by using the Latin botanical names.

Often Iannicelli's quest for a particular wood leads far afield. While on business in Madrid, he took an afternoon off to track down some olive wood. Olive trees are known to live hundreds, even thousands, of years and produce a beautifully figured wood.

This article first appeared in American Rifleman

Dr. Joseph Iannicelli proudly displays his unique collection of Browning BBR .300 Winchester Magnum rifles, Fajen-stocked with rare and exotic woods from Europe, Asia, Africa, and North and South America.

The search finally led to an eccentric wood dealer who had probably the finest assortment of wood in Madrid, but no olive. But, while poking around the man's shop, Iannicelli spied a crooked log that proved to be Cameroon ebony. After a brief haggle in French, a deal was struck and Iannicelli returned to his hotel, where a pair of doormen unloaded the black log and shunted it to a side entrance. Iannicelli recalls with a chuckle the plaintive looks on their faces as they contemplated the black stains on their white gloves.

Later, a curious but friendly Iberia airline check-in clerk overlooked the excess baggage weight contributed by the 80-pound log. A clerk in New York let it slip onto a connecting flight.

When Fajen received the log, he studied it for a week before deciding how to cut a blank from the dusky mass. He recalled that he felt like a diamond cutter. In any case, his decision was well-made. The cut blank was dried over a period of nine months and then fashioned into a jet-black stock with just a trace of gray marbling to identify its natural origin.

When a rough stock blank is first obtained, the ends are coated with a high-temperature melting wax to prevent end checking caused by excessive moisture loss from the end grain. Fajen also uses an oil-based clear finishing material to seal cross-grain areas of some imported high-density woods to help prevent seasoning checks from developing.

All blanks are conditioned to a moisture content of 8-10 percent in the dry-kiln using mild schedules. Typically, a schedule would be to use a low chamber temperature (110°F) with a wet-bulb temperature to give 85 percent relative humidity. Then the drying temperature is increased to minimize stress within the stock blank and still achieve the desired 8 percent moisture content.

Once the blank is conditioned to the proper moisture content, the shaping and inletting are started. The rough outline of the stock is cut on a bandsaw, with the contour carving being performed on a spindle duplicating carving machine.

Some of the unusual woods had unexpected

CASE 1, from left:
Sucupira,
Japanese Maple,
Pink Ivory,
Zebrawood,
Goncalo Alves,
Cocobolo.

CASE 2, from left:
Kingwood,
Pistachio,
Tulipwood,
Mango,
Ipe,
Purpleheart.

CASE 3, from left:
Redwood,
Sumac,
Honeylocust,
Tiger Myrtle,
Osage Orange,
Mesquite.

CASE 4, from left:
Shellflame Maple,
Bird's-eye Maple,
Fiddleback Maple,
English Walnut,
Black Walnut,
Claro Walnut.

GUNSTOCK WOODS OF THE WORLD

Wood Origin	Botanical Name	Density[1] (lbs./cu. ft.)	Net Weight[3] Stock	Color
NORTH AMERICAN				
Maple-Bird's-eye	Acer saccharum	44	2 lbs., 7 ozs.	light brown
Maple-Fiddleback	A. saccharum		2 lbs., 2.5 ozs.	light brown
Maple-Shellflame	A. macrophyllum	34	1 lbs., 13.5 ozs.	light brown
Walnut-Bastogne	Juglans regia/hindsii	—	2 lbs., 2.5 ozs.	dark brown
Walnut-Black	J. nigra	38	2 lbs., 7.5 ozs.	chocolate brown
Walnut-English	J. regia	33	2 lbs., 2.5 ozs.	light brown
Walnut-Claro	J. hindsii	—	2 lbs., 2 ozs.	tan brown
WESTERN UNITED STATES				
Myrtle-Tiger	Umbrellularia californica	39	1 lbs., 15.5 ozs.	light brown with darker streaks
Myrtle	U. californica		2 lbs., 2.5 ozs.	
Mesquite	Prosopis pupescens	54	2 lbs., 5 ozs.	dark golden brown
Madrone	Arbutus menziesii	45	2 lbs., 3 ozs.	light reddish brown
Redwood	Sequoia sempervirens	28	1 lbs., 13 ozs.	dark reddish brown
Osage Orange	Maclura pomifera	55	2 lbs., 14 ozs.	golden brown
Pistachio	Pisticia vera	—	3 lbs.	light to dark brown
EASTERN UNITED STATES				
Butternut	Juglans cinerea	25	1 lbs., 13.5 ozs.	light chestnut brown
Cherry	Prunus serotina	35	2 lbs., 6 ozs.	light reddish brown
Sycamore	Platanus occidentalis	35	2 lbs., 2 ozs.	light brown
Honey Locust	Gleditsia triacanthos	44	2 lbs., 8 ozs.	light reddish brown
Ash	Fraxinus americana	42	2 lbs., 3.5 ozs.	light grayish brown
Mango	Mangifera indica	—	1 lbs., 15 ozs.	light golden brown
Pecan	Carya illinoensis	46	2 lbs., 13 ozs.	light brown
AFRICA AND ASIA				
Indian Rosewood	Dalbergia latifolia	53	2 lbs., 12 ozs.	dark purplish brown
Padauk	Pterocarpus dalbergioides	40	2 lbs., 8.5 ozs.	yellowish brown with reddish streaks
Ebony	Diospyros crassiflora	73	3 lbs., 10 ozs.	black
Pink Ivory	Rhamnus zeyneri	62	3 lbs., 6.5 ozs.	bright pink
Zebrawood	Microberlinia brazzavillensis	46	2 lbs., 6 ozs.	light brown with dark brown stripes
Teak	Tectona grandis	38	2 lbs., 6 ozs.	brown with darker streak
Yama	Acer palmatum	30	1 lbs., 11.5 ozs.	light grayish brown
SOUTH AND CENTRAL AMERICA				
Ipe	Tabebuia serratifolia	68	3 lbs., 2 ozs.	dark olive brown
Sucupira	Bowdichia spp.	62	3 lbs., 5 ozs.	chocolate brown
Goncalo Alves	Astronium fraxinifollum	66	3 lbs., 3.5 ozs.	golden brown with blackish brown streak
Cocobolo	Dalbergia hypoleuca	68	3 lbs., 10 ozs.	variegated with orange and dark reddish brown streaks
Tulipwood	Dalbergia frutescens var. tomentosa	76	3 lbs., 5 ozs.	light yellowish brown with stripes of red
Purpleheart	Petogyne paniculate	54	2 lbs., 12 ozs.	deep purple to brown
Kingwood	Dalbergia cearensis	75	2 lbs., 9 ozs.	variable dark reddish, violet brown

[1]Weight based on 12% moisture content, with data from Wood Handbook. No. 72, USDA Forest Service, Forest Products Laboratory, Madison, Wisconsin: [2]Browning barreled action weighed 6 lbs.; [3]Fajen Aristocrat stock design.

effects when Fajen workers began to shape them into stocks. Some had to be worked with carbon-steel tool bits that wear faster than carbide-steel-tipped bits. The carbon-steel bits, though, can be frequently sharpened to maintain precision in in-letting. The high-density woods like cocobolo, ebony, and tulip-wood required that bits be sharpened again and again. Woods like ipe, sucupira, and teak also contain silica, which has an additional dulling effect on tools.

Some of the woods, like Indian rosewood, padauk, and teak, contain oils that clog files and sandpaper, requiring frequent cleaning of files and replacement of paper.

Shaping some of the stocks was tough not just on tools but on the five Fajen craftsmen who put an average of 23 hours' labor into each stock. Workers found that the sawdust of cocobolo and goncalo alves was particularly irritating to skin and nasal passages.

Once the stock has been given a pass with 400-grit abrasive paper, it is ready for the final step—the application of the decorative and protective finish. Some of the stocks are stained or filled to enhance appearance. For example, the pores of honeylocust were filled with a dark filler substance. The filled pores were a nice contrast to the lighter-colored honeylocust wood. All the maple stocks were stained with a light brown water-based stain to enhance the figure.

Next, a thin film of boiled linseed oil, only in the amount that will be absorbed by the wood, is applied and allowed to dry at room temperature for three or four days. Then five successive coats of Fullerplast catalyzed clear polyester resin are sprayed on. Each coat is allowed to become "tacky" before the following coat is applied. The finish is then allowed to dry and cure for about a week.

Next, the stock is scuff-sanded using 200-grit silicon carbide paper to obtain a level, uniform surface on the stock, and another five coats of polyester finish are sprayed on in the same manner described above. After the finish dries, it is scuff-sanded again, this time with successive grades of abrasive paper, finishing with 600-grit. Finally, the stock is polished, using a fine-grade buffing compound.

The result is a beautifully smooth and shiny finish that enhances the color and grain of the wood.

The woods represented run the gamut of color, density, weight and texture. The color variations run from the very light maples to pink (madrone) to dark chocolate brown (sucupira and walnuts) to ebony.

Texture is another wood property where there is great variation—the diffuse porous nature of pistachio and sycamore contrasts with the ringed porous nature of honeylocust and osage orange. The latter woods are especially hard to process.

Finally, there is a wide variance in the density of the Iannicelli woods Redwood, for example, weighs just 28 pounds per cubic foot, while tulipwood tips the scales at a hefty 76 pounds per cubic foot.

While all the stocks in the Iannicelli collection are magnificent, two stand out above the others. The first is the redwood stock fashioned from a cross-sectional piece of burl. That's right—a cross-sectional piece! The growth rings are perpendicular to the stock, with the center pith of the log near the center of the stock.

The strength properties of the wood are weakest along the growth rings, and I cannot recall ever seeing a stock made with the grain so oriented. In order to ensure proper stock strength, maple rods were glued into several holes drilled lengthwise in the blank to give strength across the growth rings of the wood.

But the real showstopper of the Iannicelli collection is the African pink ivory. It really is pink, almost the shade of bubble gum. Legend has it that only the sons of Zulu chiefs are permitted to cut this rare wood species. When a chief's son is able to fell a tree and make a spear from the wood, he is considered to have reached manhood. Any other member of the tribe who cuts or possesses the wood is subject to death.

The wood from the pink ivory tree is very difficult to obtain because of its limited growing range and the small number of trees available for cutting. The average size of a mature tree is 20″ in diameter. As a timber species, it would not be a commercial proposition. But pink ivory, because of its color and rarity, has become a much sought-after wood for speciality items like custom knife handles. Getting a piece big enough for a rifle stock was one of Iannicelli's greatest wood-gathering feats.

There are over 50 different woods currently in the collection, and Iannicelli is advancing toward his goal of 100. During the past year, Fajen has completed stocks of chestnut, magnolia, sumac, bubinga, greenheart, koa, Macassar ebony and monkey pod, among others. Other woods in progress include lignum vitae, olive, wenge, blackbean, Cooktown ironwood, as well as some woods from the state of Georgia—dogwood, eastern red cedar, holly, and persimmon. His want list includes exotics such as black mangrove, bristlecone pine, English yew, ginkgo, mulga, satinwood, and snakewood.

Iannicelli's wife Betty describes the collection as his "magnificent obsession." It's hard to imagine a better marriage of function and art, or a better way to display the beauty and variety of natural wood.

Checkering, from Crude to Baroque

Jim Carmichel

It is my habit to keep a beautiful gun or two on a nearby chair when I'm working in my office. Leaning back in my ancient leather desk chair, I often glide open the action of a rifle or kiss a shotgun to my cheek for the same reasons that others light a pipe or brew a pot of English tea. Fine guns give light and warmth to the coldest, darkest hours and provide drink and nourishment to a parched and empty spirit.

The guns are also nice to talk about when my hunting pals drop by. Even better are the times when non-shooters find themsleves in my office and are forced to lay hands on the fearsome things if they want a seat.

"Go ahead and look it over," I always tell them, watching their hands spring from cold steel to

This article first appeared in Outdoor Life

Stanley W. Trzoniec photo

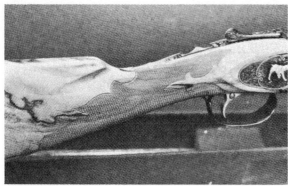

Style of this Pachmayr checkering on Model 21 Winchester is like late baroque music, so sophisticated it's almost decadent. Could you bring yourself to go on a really rough grouse hunt with it?

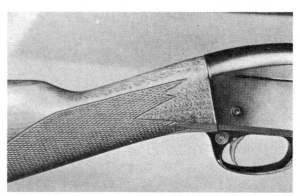

This machine-cut checkering is "borderless," though it is actually edged with what can only be called a gutter. Machine cutting is accurate and quite sharp on this example, but real border would dress it up.

During the checkering renaissance, basketweave style came into considerable vogue. Some fanciers find it too ostentatious; others feel that it's easy to do with precision because some lines are skipped.

Impressed checkering was done with hot dies forced into wood. That saved expensive labor, but most aficionados think it's hideous. You may recognize this Winchester pattern; they did it better than others.

warm wood and their eyes dart over the complex contours like the eyes of a field mouse that has strayed too far from its nest and is startled by the strange terrain.

Almost always, their attention comes to rest on the ornate checkering pattern, and I'm asked to describe its purpose.

FUNCTION AND ART

"It's called checkering." I explain, "and it's just a nonslip surface so that the shooter can get a better grip, especially if he has wet hands or is wearing gloves."

"But why is it so fancy?" they ask.

"Because checkering has evolved into some-

thing of an art form, similiar to decorative carving or engraving on metal. It's the old business of art following function, I suppose."

But then I ask myself what really happened. How could checkering on gunstocks, which has a function no more glamorous than sand on an icy sidewalk, become the focal point of custom guns costing many hundreds or thousands of dollars? When custom rifles and shotguns are displayed at large exhibitions such as the one at the National Rifle Association's annual convention, throngs of visitors scrutinize the lavish checkering patterns like gourmets considering princely banquet fare. They often compare one craftsman with another by considering the ornateness of the designs, the delicacy of each diamond, and the virginal primness of the borders.

To many of these people, it doesn't seem to matter that the rifle may not be accurate enough to hit a haystack.

Newcomers to the shooting sports often wonder why reports on new guns by firearms writers make such a fuss about checkering. After all, checkering is just checkering, and who cares what it looks like so long as it does its job?

That's pretty much how checkering is viewed in Europe. A snazzy Italian gun builder, for example, will blissfully invest weeks of time and wages in engraving every square millimeter of exposed steel but won't even discuss a respectable checkering job. In fact, it's common for some big-name Continental gunmakers to farm out checkering to local folks on a piecework basis. Like any piecework, the faster it's done, the better. I recently inspected a $12,000 Italian shotgun with checkering that looked as though it had been executed by a blind rooster scratching for a worm. Even the checkering on the vaunted London "best" shotguns, while neat and workmanlike, tends to be as plain and unremarkable as a potato skin. That's why it's not unheard of for American sportsmen, who shell out big bucks for European shooting ware, to specify that the wood be left uncheckered. They prefer to have the job done to perfection by a craftsman in this country. To be sure, some superlative checkering is done in Europe, but mostly on special order only.

Here in the United States, most really great checkering is seen on custom rifles. The reason for this is simply that most of our top craftsmen specialize in rifle work. When they do turn their attention to custom stocks for shotguns, the results are no less impressive.

CHECKERING'S HISTORY

Checkering as a form of firearms ornamentation goes back hundreds of years. The earliest surviving examples of checkering, in fact, suggest that its purpose was more decorative than utilitarian. By the 18th century, checkering was commonplace on English fowling pieces, and it was occasionally found on German Jaeger rifles and even American long rifles. Sometimes, the ornamental effect was accented by driving brass or silver tacks into the pattern. The best examples of checkering from that era were found, predictably, on British dueling pistols, thereby emphasizing an English gentlemen's determination to get a firm grip on serious matters.

Guns made during the middle of the 19th century were scarcely checkered if at all because the emphasis of the times had shifted to firepower. Checkering on a lever-action Winchester, for example, was a special-order luxury. By the end of the century, however, American manufacturers were rediscovering checkering and using it as it had never been used before. Highgrade Parker, Fox, and Ithaca shotguns, to mention but a few, were endowed with checkering patterns of unprecedented elaborateness. The gripping surfaces of sporting firearms of similar quality from the gun-building shops of Belgium and Germany often had gripping surfaces embellished with some form of carving, such as oak leaves, instead of checkering. Fancy checkering came to have a distinctly American flavor. Perhaps this is exactly

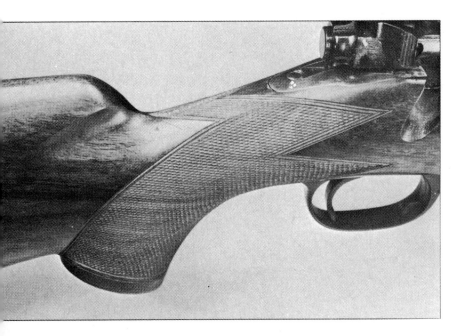

Old-fashioned, simple checkering done by Griffin & Howe is decorative and utilitarian. Nicely cut, fairly wide border enhances simple point pattern.

what our shotgun makers wanted to achieve, maybe to give them an edge because American smoothbores were relentlessly compared with their European counterparts. Though the great double-barreled shotguns that were made at the beginning of the 20th century are no longer in production, there is one remaining fine American double, the Winchester Model 21. It's checkered in a fancy pattern reminiscent of the Gilded Age.

During the 1930s. checkering on both rifle and shotgun stocks became what reason, utility, and economics said it ought to be. A very good example of the checkering of that time is to be found on older Model 70 Winchester rifles. The pattern is rather plain but ample, the work neatly done and framed by a simple but pleasing hand-cut border. Even the custom rifles of that time, as typified by the stylish stocks made by Griffin & Howe and the Sedgley Rifle Company, featured checkering that did its job without trying to steal the show.

How, then, did our current mania for fancy checkering come about?

Like so many other things that have changed our ways, fancy checkering was a fallout from World War II. The hundreds of thousands of liberated rifles that were brought home by GIs created a hurricane of interest in do-it-yourself custom rifles. Hobby gunsmiths, who sporterized their Mausers with semi-finished stocks, were usually so pleased with the result that they often customized additional rifles. After a while, these do-it-yourselfers took the big step and did a checkering job. Even if the results were not satisfactory—frequently they were not—anyone who tried his hand with a checkering tool couldn't help but gain a lot of respect for good checkering. The patient souls who developed skill with checkering tools often attempted to show off their newfound talent by executing increasingly difficult patterns or tricky-looking layouts. Skip-line checkering (also known as French checkering or basketweave) became the rage and was much copied after it appeared on the dazzling Weatherby magnum rifles.

About the same time, the stars of custom stockmakers were beginning to glow more brightly in the shooting heavens. There had always been makers of custom guns, and good ones, too, but they never achieved celebrity status until my sainted predecessor as Shooting Editor, Jack O'-Connor, sang their praises in the pages of OUTDOOR LIFE. Old Jack loved fine checkering and extolled the virtues of a steady-handed craftsman who could create delicate rows of diamonds and cut a line of checkering right up to the edge of the pattern without causing a nick or runover and therefore did not need a border to wipe out mistakes.

Borderless checkering came into vogue and served as a convenient showcase for a stockmaker's talents. Of course, lost of borderless checkering wasn't good checkering, and more than a few jobs would have benefited by the addition of a nicely cut border to erase the runovers or to dress up an otherwise plain pattern.

Just about the time the craze for fine checkering ws in full flower, some major gun manufacturers switched from traditional hand checkering to a machine process that pressed pseudo-checkering into the wood. Their timing could not have been worse. Gun lovers, especially those who had only recently come to appreciate good checkering, reacted like poleaxed oxen. It was ugly. Ugly! Merely hearing the phrase "pressed checkering" made the blood of gun fanciers run hot and vengeful. Of all the criticisms that were heaped on the new Model 70 Winchester of 1964, none was more scornful than those directed at its pressed checkering. Ironically, Winchester did the best of the pressed checkering.

Bill Ruger capitalized on the gun-buying public's negative reactions to pressed checkering by offering rifles that featured handcut checkering. Actually, the checkering on Ruger guns was and is cut by hand-held, hand controlled electric tools that work like miniature buzz saws. The same equipment is used by many custom stockmakers, though many craftsmen used true hand tools, Ruger's craftsmen (most of them are crafts*women*) were trained by the late Len Brownell, himself a top stockmaker, and have demonstrated that a stock can be tastefully checkered by hand with these electric tools about as quickly as it can be done by complex cutting machines. The Ruger checkering department is really something to see, what with the checkering tools buzzing like angry bees and the women smoking cigarettes like fiends.

Happily, the gun industry has been able to economically replace pressed checkering with computer-controlled machinery that cuts a pattern that is difficult to distinguish from the handcut variety. This is the way most American checkering is now done. You've probable noticed that some manufacturers' brochures proclaim "cut" and "borderless" checkering. That is, perhaps, the final irony because most of today's machine-cut point patterns would look considerably better if they were framed by a nice border. This is especially true of those patterns that are edged (not bordered) by what can best be discribed as a routed gutter.

On the custom front, some checkering has become so fancy and delicate that one is hesitant about touching it, much less using it as a gripping surface when shooting. But it sure is mighty pretty to look at.

PART SEVEN

IMPROVEMENTS, ALTERATIONS, AND GUNSMITHING

Practical Tips on Shotgun Fit

Hugh Birnbaum

One of the great truths of shotgunning is that a gun that fits well nearly always hits well, assuming reasonable shooter competence. The reverse is also true. A gun that fits badly is likely to cost you lost targets or game, and it may add injury to insult by subjecting you to cruel and unusual punishment every time you pull the trigger.

Fortunately, fitting a shotgun properly doesn't require divine revelation. It is mainly an exercise in logic applied in small increments. Incidentally, many riflemen would benefit by paying more attention to gun fit than they do. Several criteria of good shotgun fit apply to rifles as well.

Before examining the key components of good long-gun fit, it's helpful to establish a few ground rules, First, as Polonius counseled, "To thine own self be true." Stock shapes and dimensions that suit one shooter may be totally wrong for another. So don't start with the intention of duplicating the gun setup that works wonders for the club hotshot. Unless you are identical twins, it propably won't work for you.

Second, tailor your gun to fit your primary needs. A shotgun that performs spectacularly on the trap field may be a disaster in the game field, and vice versa. If you have several shooting interests that are incompatible from the standpoint of gun fit, it is usually not a good idea to try to achieve a compromise solution. You'll end up with a gun that isn't quite right for any one application. In ascending order of cost, you can tackle the problem by: optimizing the fit for your main interest and taking your lumps in secondary areas; purchasing and fitting separate stocks for different purposes, switching stocks as necessary; or, budget permitting, buying one or more additional shotguns suited to discrete needs.

Third, don't expect miracles. A shotgun or rifle that fits well will be more comfortable to shoot because it minimizes strain, hence fatigue, and, in the case of hard-kicking guns reduces felt recoil appreciably. A properly fitted shotgun will be easier to swing smoothly at a rate appropriate to the type of shooting. A rifle that fits well will

This article first appeared in American Rifleman

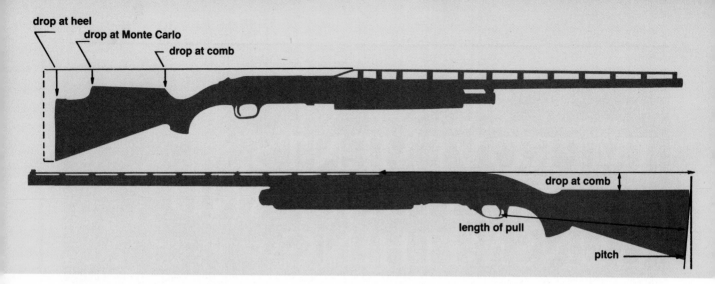

These drawings illustrate important aspects of stock design that affect gun fit. Shotgun in upper drawing, combining high rib and high Monte Carlo comb, has sophisticated contours that can be of great benefit to advanced trapshooters but would actually be detrimental in typical field situations. Desirable measurements and contours depend on gun's purpose plus—most importantly—owner's physique and shooting style. By modifying stock, you may be able to improve your shooting.

facilitate rapid, consistent eye placement in relation to sights or scope and will be easier to hold on target.

The net result of a properly fitting gun is usually a noticeable improvement in your score with targets or game. Nonetheless, fitting your gun properly won't magically confer shooting skills you didn't have before. It will simply allow you to make better use of the skills you have or remove obstacles to developing your ability more fully.

Fourth, don't expect to take definitive measurements directly from a try stock that isn't fitted to your gun or an absolutely identical gun. A stock configuration that works well for you with a Remington Model 870 up front may not work at all well for you with a Model 1100 or Model 3200 because of differences in weight, balance, and various dimensions. A try stock can be extremely useful for getting into the ball park, but there is no substitute for cut-and-try fitting with the actual gun.

Fifth, and very important, don't overlook the possibility that your gun fits just fine as is. Trapshooters in particular tend to tinker more than they probably should with gun fit at the first sign of dipping scores. Sometimes reviewing basic shooting technique is all that's needed to end a slump. Every competitive shooter should probably paste a big sticker on his gunstock bearing the admonition: "Don't fix what ain't broke!" With these caveats in mind, here are factors to consider in fitting a shotgun.

For most shooters, the length of pull is the most critical dimension. It is also the easiest to modify. The length of pull is the distance from the middle of the front surface of the trigger to the middle of the rear surface of the buttplate or recoil pad. When the pull length is too long, it is difficult to raise the gun to shooting position without snagging the butt, the gun may feel excessively muzzle-heavy and difficult to move, and you will find your lateral swing limited.

A right-shoulder shooter will be most hampered when swinging to the right, and a lefty will jam up tight swinging left. The obvious remedy is to shorten the stock. The tricky part is deciding how much to shorten it. Before grasping that nettle, consider the opposite problem.

A stock that is too short betrays its shortcoming two ways, one obvious and one sometimes quite subtle. The obvious sign is that recoil rams the base joint of the thumb of the trigger hand against the nose, cheek, or shooting glasses. This is hard to ignore and begs for remedial action. The subtle sign is a persistent tendency to move the gun too fast. If you find yourself often sweeping past the target before you can react, or if you consistently overlead targets, suspect that the stock may be too short. Don't jump to this conclusion too quickly, though, because these problems can occur with a shotgun that fits correctly but is too muzzle-light for the type of shooting or your energy level.

If your gun accepts interchangeable barrels, try a longer tube or barrel set, if available, before altering a stock that feels good in other respects. Installing weights forward of the balance point can also correct a muzzle-light condition. If more weight up front doesn't help, then consider lengthening the stock. Again we come to the tricky question: How much?

Whether you're shortening a stock that's too long or lengthening one that's too short, the ultimate goal is a stock short enough to mount smoothly when raising the gun, held horizontally, from the waist, short enough to swing easily, yet long enough to keep your thumb joint from attacking you under recoil. For most people, there is no single magic length required to accomplish these modest aims. You can probably find happiness with a stock that deviates ± ¼-inch from the ideal.

The easiest way to experiment with stock lenght is by trying recoil pads of different thicknesses. Many gun stores stock recoil pads ranging from little more than ¼-inch to as much as 1¼-inches thick. You can also buy composition spacers to sandwich between stock and recoil pad to increase overall length in increments of ⅛-inch or so. To save time and money, ignore aesthetics and install or remove pads and spacers yourself without matching them to the stock contours. When you find the length that suits you best, then have a gunsmith do a neat installation job.

If you wish to shorten a stock so much that you must remove wood, I strongly recommend having a skilled gunsmith do it. It's distressingly easy to ruin a good stock with inept hacking. For major lengthening of a stock, it is usually possible to add an insert contoured to follow the lines of the basic wood. Doing this well isn't a job for the casual tinkerer.

In all cases of adjusting stock length or other important dimensions, shoot the gun before final finishing. The time to find out that you want the stock a shade longer or shorter or higher or lower is before you've run up a bill restoring the wood to its former cosmetic glory.

If you live in a region where the temperature changes markedly with the seasons, remember that the length of pull that is perfect when you wear a light-weight shooting vest may feel impossibly long when you're bundled in bulky winter clothes. One practical solution is to fit the butt with two recoil pads: a thick one for the summer and a thinner one for cold weather.

If you perfer to keep the stock the same year-round, try increasing the amount of padding under the gun patch of your summer shooting vest to match the total thickness of material where you seat the recoil pad over winter wear. I've tried both methods and they both work.

Shooters with unusually large or small hands may find that the wrist of the stock, the portion grasped by the trigger hand, can be a problem area. If the wrist is too slender, build it up to fit your hand with layers of tape, epoxy, or wood. Tape is cheap and easily removed. Epoxy is cheap and is permanent for all practical purposes. Adding wood is costly if it is done artfully, but is less likely to ruin the appearance of the stock. Thinning a wrist that is too thick must be done carefully. Remove the least amount of wood to achieve a comfortable, secure grip. Removing too much wood will weaken the wrist and leave it more susceptible to splitting or shattering under heavy recoil.

Competitive shotgunners (and riflemen) may find that the curve of a full pistol grip fails to position the hand properly for optimum trigger control. The same methods suggested for adjusting wrist thickness may be used to shorten or lengthen the distance from the front of the pistol grip to the trigger. If you build up the grip to move your hand nearer the trigger, be sure to

Stock's wrist—its pistol-grip area—may be ideal for one gunner and too slender for another. You might think that only ham-handed gunners need thick, hand-filling pistol grip, but frequently small-fisted shooters require added support here to move fingers forward, thus putting trigger finger in proper position and exerting firm, positive gun control. Artfully executed wood extensions are unobtrusive or handsome—but costly. Epoxy build-ups are less attractive but cheap and effective.

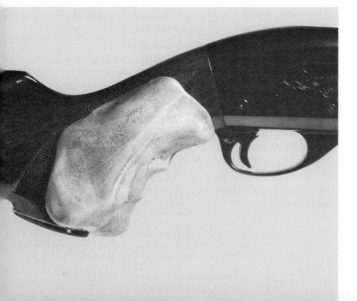

leave enough room behind the trigger guard. If you don't, it will strike your middle finger when the gun recoils.

Because a shotgunner's <u>master eye functions</u> <u>as a rear sight,</u> the eye's position relative to the rib and front bead determines the center of impact. If the eye position is high, the shot charge will go high. If the eye position is low, the pattern will go low. Similarly, lateral displacement of the eye relative to the rib will affect corresponding sideward shifting of the center of impact. To shoot a shotgun consistently, then, you've got to be able to position your eye consistently and repeatably every time you mount the gun.

The stock dimension that affects consistent eye placement is drop. Drop refers to the vertical distance from the top edge of the comb of the stock to the extended sighting plane of the rib. If the stock comb parallels the rib, one measurement tells all. If the comb slopes up or down with respect to the rib, several measurements are required to describe the relationship adequately. Most commonly they are drop at the nose, or point, of the comb (the frontmost part) and drop at the heel of the comb (the rearmost part). With a standard straight stock, the heel of the comb is also the heel of the stock, and the measurement may be identified as drop at heel, with no mention of the word "comb." With a Monte Carlo stock, drop at heel of comb and at heel of stock will be given separately.

There is no pat answer as to proper comb height. It depends on how you mount the gun, the configuration of your face, the type of shooting you do and your personal preference as to pattern placement relative to the line of sight. For most shotgunners, the comb is the right height when it places the eye in the individually required position time after time to deliver the shot charge on target.

If you consistently <u>shoot too high,</u> <u>lower</u> the comb until the pattern is where you want it. If you frequently shoot too low, build up the comb height. For a right-handed shooter, if you often shoot to the left of the desired point, thin the comb by reducing the left side until the pattern centers. If you're often blasting air to the right of the target, build up the left side of the comb to crank in a bit of left windage. These comments assume that the problems described are attributable to misplacement of the shooting eye rather than to a bent or otherwise defective barrel.

As a rule, trap guns have relatively high combs to provide built-in vertical lead appropriate to shooting at rising targets. Skeet and field guns generally have lower combs and shoot flatter, the better to cope with targets that are not necessarily climbing.

A quick check on comb height is simply to

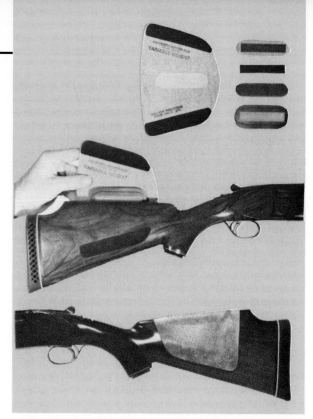

Adjustable and removable pad can raise comb or alter its slope. Purpose of this temporary modification may be to make gun more versatile, improve alignment of shooter's eye, or relieve cheek-bruising recoil—or all three desirable factors.

mount the gun with your eyes closed, holding it and cheeking it in your accustomed manner. Open your eyes and glance along the rib. If the comb is right for you, you should be looking right down the centerline of the rib, with little of the upper surface of the rib visible. If the gun has a middle bead as well as a front bead, the latter should appear to be perched just atop the former in a figure eight. This condition represents classic, textbook perfection. In fact, many shooters prefer to see a bit more or less rib, but that doesn't invalidate the concept.

Comb height aside, the specific shape and slope, if any, of the comb can make the difference between shooting a gun with great pleasure or with considerable trepidation. If the comb rises up and smites you mightily on the cheek every time you pull the trigger, whatever virtues the gun may have will soon be obscured by the flinch you develop in anticipation of that blow.

As a rule, a comb that slopes upward from heel to nose, as is often found on field and skeet guns, is more likely to savage the cheek than a comb that is reasonably level or that slopes downward from heel to nose. The reason is that recoil drives an upward-sloping comb against the cheek like a wedge, whereas a level comb slides by benignly and a down-sloping comb moves away from the cheek under recoil.

Note that this is not an across-the-board indictment of field and skeet stocks. It is simply a clue as to how you may be able to stop a cherished long gun from beating you more senseless than you care to be. If your cheek is being battered, check the comb height and slope first.

Altering comb height is easy if you're not obsessed with aesthetics. Judicious rasping can lower a comb, change its slope, or thin it as necessary. Remove small amounts of wood and test-fire the gun after each alteration, firing at clay targets or a pattern board. Don't try to judge by appearances. What looks right won't necessarily feel or shoot right. When you've reduced the dimensions sufficiently, refinish the stock yourself or have it done professionally.

Building up comb height is easier still. Meadow Industries' stick-on Convert-A-Stock pads, available through gun dealers and the "pro shops" of many trap and skeet clubs, permit experimenting with different comb heights and slopes without disfiguring wood or spending gobs of money. When you find a combination of height and slope that works for you, leave the Convert-A-Stock sandwich on the stock or have a gunsmith translate it into a nicely matched wooden addition to the stock. On a less presentable but fully functional level, you can build up a comb with layers of duct tape, moleskin, or any other material smooth enough not to abrade your cheek. A trapshooter I know once performed emergency surgery on a new and super-expensive competition gun at the range by taping sections cut from a styrofoam coffee cup onto the comb until he felt he was smoking claybirds satisfactorily. It looked bizarre, but it worked just fine. When fitting a stock with a sloping comb, establish the length of pull to your satisfaction before tackling comb height. With a typical field stock, for example,

with less drop at the nose than at the heel, shortening the pull will place your cheek farther forward, where the comb is higher, and lengthening the pull will position your cheek more to the rear, where the comb is lower. Any attempt to regulate comb height before zeroing in on the length of pull would therefore be futile.

Occasionally, a shotgunner finds that he has no problem lining up for a good view over the rib, except that the recoil pad or buttplate ends up too high or low in relation to the shoulder pocket where it should ideally be seated. This may allow the gun to slip out of position or may cause discomfort because too much recoil energy is transferred to the shooter over too small an area. The dimensions that are normally altered to dehorn this dilemma are the drop at the heel and toe of the stock. In simplest terms, this entails shifting the recoil pad up or down relative to the gun butt until the pad nestles comfortably in the shoulder pocket at the same time you achieve proper cheek contact with the comb while sighting along the rib.

The quick-and-dirty solution is to drill new holes for the recoil-pad screws that allow you to reposition it where it does the most good. A neater approach is to fit the stock with a Morgan Adjustable Recoil Pad, which lets you shift the movable pad up or down over an ample range and lock it in the position you prefer with an ordinary screwdriver. If you change your mind or shooting technique, just loosen the screw and try a different setting. Morgan pads are easier to find at gunsmithing establishments and retailers catering to trap and skeet shooters than at less specialized emporia.

Pitch is an aspect of fitting a shotgun or rifle that is often neglected but can strongly influence the manner in which the shooter experiences re-

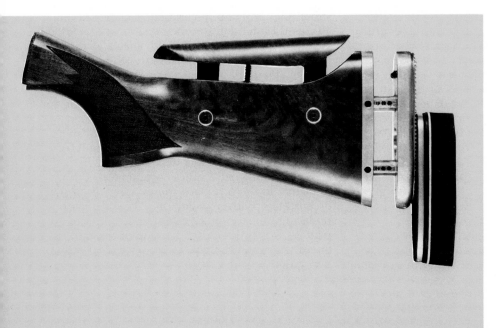

Adjustable stock like Fajen's—obviously inspired by try stocks that custom gunmakers employ to determine individual fit—can accommodate almost any physique. But ordinary field-grade stocks can be altered as well.

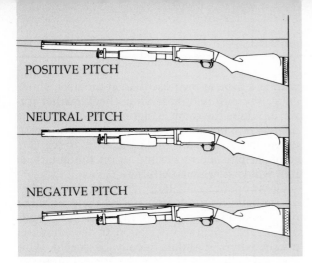

Stock's pitch is angle between bore's axis and plane of buttplate or recoil pad. With butt held vertically and firmly against any straight wall, note the angle of the bore's axis. Neutral pitch is 90°. Negative pitch, or pitch down, is less than 90° and positive pitch, or pitch up, is more than 90°.

Recoil pad or butt extension can increase length of pull, and adjustable type can alter butt position against shoulder as well as comb position against cheek—raising or lowering stock for better eye alignment and/or reduction of perceived recoil. Often this can be done without altering pitch, cast, or slope of comb.

coil. Pitch refers to the angle included between the principal plane of the rear surface of the recoil pad or buttplate and the bore axis. Neutral pitch exists when that angle is 90 degrees. When the angle is less than 90 degrees, the gun exhibits pitch down, or negative pitch. When the angle is greater than 90 degrees, the condition is termed pitch up, or positive pitch.

You can visualize this by placing the gun butt solidly against a wall and noting whether the barrel axis is level or aimed above or below the horizontal. Pitch is often expressed as the distance the muzzle rises above or descends below the horizontal. That can be misleading, however, as an inch of muzzle displacement for a 34-inch barrel, for example, describes a significantly different angle of pitch than an inch of muzzle displacement for a 26-inch barrel. If you ever have to quantify pitch for a custom stockmaker, you can eliminate ambiguity by providing an angular measurement relative to the horizontal. Five degrees pitch up is the same for any barrel length.

Pitch is important to shooting comfort because it dictates the way the recoil pad or buttplate seats against your shoulder. Excessive pitch up or down can cause the toe or heel of the pad to press sharply against the pectoral or shoulder area, acting like a battering ram during recoil. Extreme down pitch also tends to wedge the stock upward during recoil, driving the comb painfully into the cheek. The pitch is right for you when the recoil pad or buttplate bears evenly against your shoulder pocket with the gun mounted. You should definitely not feel the heel or toe digging into your anatomy.

Minor adjustments in the pitch are easy to improvise by shimming the recoil pad or buttplate away from the stock at heel or toe with bits of paper, thin card stock, or any other materials that don't compress. Such shimming may create an unsightly gap between stock and pad. On the other hand, it's free. Any competent gunsmith will be happy to do a beautiful job for you. It won't be free. Major changes in pitch, which are rarely required, are best left to professionals.

When refining the fit of a stock, give some thought to the width of the wood and to the shape of the recoil pad or buttplate. Up forward, decide whether or not you can grasp the fore-end comfortably. If it is too skinny, you may be able to replace it with a bulkier style or build it up with additional wood. A fore-end that is too large for you to grasp securely may be slimmed down through replacement or rasping. Replacing a fore-end does not necessarily entail custom work and prices.

Some shotguns are produced in multiple variations stocked differently for trap, skeet, field shooting, or live-bird competition. Occasionally a fore-end intended for one version will be fatter or slimmer than one supplied normally for another configuration. When that is the case, you may be able to buy a replacement off the shelf at a reasonable price, especially if you're not overly concerned with matching its tone and figure to the gun's buttstock.

A buttstock that is too narrow where it meets your shoulder will intensify recoil sensation compared to one that makes wider contact. If that's a problem for you, a quick, economical, but unattractive fix is to replace the existing recoil pad or buttplate with a wider one. The alternative, replacing the stock with a beefier custom design, may be worth the cost if you shoot the gun frequently and intensively, as occurs with competition guns.

Recoil pads and buttplates are often taken for granted or, worse yet, selected on the basis of appearance. They merit more attention. Functionally, a pad or plate with a relatively straight profile is desirable for a gun that must be mounted and fired quickly, as in field or international-style skeet shooting. A deeply curved pad contoured to hug the shoulder, in contrast, may be just the ticket to encourage deliberate, precise gun mounting, as in American trap.

A smooth-surfaced pad or plate can slip into place rapidly without snagging clothing, a plus in the duck blind, but it may slip out of position easily, a minus on the trap field. Heavily textured pads hold position securely but can impede quick gun mounting.

One pet peeve I've harbored regarding the shoulder end of many hard-kicking shotguns and rifles is their manufacturers' persistence in fitting them with wretched metal, composition, or petrified rubber plates or alleged pads that in fact appear designed to amplify recoil rather than protect against it. In my opinion, the first modification to the stock of any shoulder arm that generates perceptible recoil should be to replace an unyielding pad or buttplate with one that provides a decent degree of cushioning.

Another incomprehensible aspect of many otherwise satisfactory recoil pads is the abrupt edges they sport. If your gun is equipped with one of these sharpies, round off the hostile edges by sanding. Your shoulder will thank you.

Nearly all mass-produced U.S.-made long guns have stocks that line up almost perfectly with the bore axis when viewed from above. Most shooters of more or less average build have no problem achieving decent alignment with this arrangement. Shooters with other than average proportions or who may have to cope with physical difficulties wrought by illness or injury, however, may find it difficult or impossible to shoot comfortably with a conventionally centered stock. For them, a stock with cast on or cast off may be necessary. The cast of a stock is the orientation of its center line, as viewed from above, relative to the bore axis. Cast off signifies displacement of the stock toward the right. Cast on refers to stock displacement toward the left. A stock with zero cast is straight. When cast is present, it is generally fairly subtle. For example, a barrel-chested trapshooter might benefit from a bit of cast off to bring the rib into better alignment with his master eye. Very large amounts of cast off have been used successfully to permit strongly right-handed shooters to fire from the right shoulder even when circumstances oblige them to aim with the left eye. Left-handers have benefited similarly in reverse.

Small amounts of cast can be introduced easily, if not handsomely, by remounting the recoil pad or buttplate slightly off the vertical center of the butt. Slight to moderate amounts of cast may be introduced by bending an existing standard stock, a process best left to expert stock specialists. If you need a stock with lots of cast, you have no choice but to have it made from scratch.

Choose your stockmaker carefully, because designing and making a stock with appreciable cast requires excellent judgment in addition to the usual skills. If the dimensions are botched, a stock with heavy cast-off will beat you half to death with recoil sensations you wouldn't believe possible.

As indicated above, much of the tweaking

Many stocks—especially factory stocks—have zero cast, meaning they extend straight back from bore's axis, as seen from above or below. Except when tailored to neutralize a shooter's physical abnormality, cast is usually very slight, its purpose being merely to improve alignment of gunner's master eye. Cast off means stock displacement toward the right; cast on angles stock leftward.

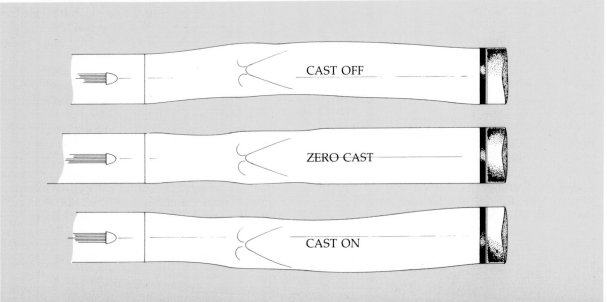

Shotgunners in these two photos demonstrate typical differences in shooting styles. Shooter at left leans forward slightly, while man at right has more erect stance and holds elbow considerably higher. Such differences strongly affect shotgun fit—height of comb, length of pull, drop at heel, and so on. Even if shooters are of same height and build, their gun-fit requirements may differ substantially, requiring stock modifications.

done to optimize a stock for a particular shooter can be performed on a do-it-yourself basis. Before you sprint to the toolbox or workshop, though, consider carefully the implications of what you are about to do. There is probably no reason not to try your hand at modifying a run-of-the-mill, factory-issue stock that may already exhibit a goodly number of dings and scratches.

But I would personally think long and hard before practicing kitchen-table surgery on a high-quality, finely finished piece of wood that represents a major portion of a gun's value. In fact, if I were thinking of altering a truly expensive stock, I would remove it from the gun and replace it with an economy-grade of stock, perhaps a semi-finished one, of the desired configuration. I would then cut-and-try to my heart's content until the cheap stock fit me perfectly. At that point, I would deliver the gun with the modified stock still attached to it, along with the pristine, high-grade handle, to a stockmaker I trusted implicitly. I would tell him to alter the expensive stock to fit like the 10-thumbs version I had cob-

bled up. That is the gun-fitting equivalent of having your cake and eating it, too.

The above comments assume you have a stock you want to improve. If you're starting clean, buying a new gun or a new stock for a gun you already own, there are other avenues you may wish to explore. For example, if cost won't be too great a deterrent, consider consulting a custom stockmaker who can design a stock for you that will fit as well as a Saville Row suit.

Less costly but still pricey, the Fajen Adjustable Trap Stock (or one of its non-Fajen clones) will let you customize the fit of a trap gun to a fare-thee-well with regard to all important dimensions. It is available in models to bolt directly onto a variety of widely used trap guns.

Of course, before embarking on an extensive and possibly expensive program of gun fitting, I would ask myself the historic question dating from the days of World World II, "Is this trip really necessary?" And before answering, I would reread the sticker on the side of the old stock that says, "Don't fix what ain't broke!"

Accurizing Springfield Armory's .45 Auto

Chick Blood

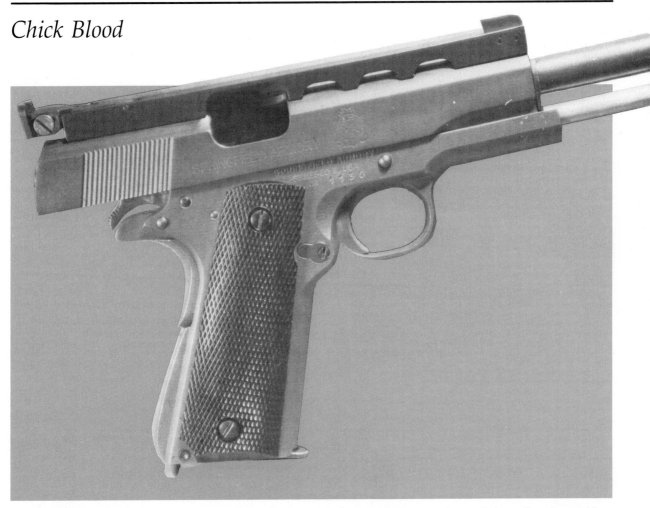

Here is author's accurized pistol, with Wichita rib. Note relieved ejection port and extended recoil spring guide.

In 1985, suggested retail prices were bad news for bull's-eye shooters with budgets. A Colt Series 80 Government Model with fixed sights listed a tad under $500, a Colt Commander topped $500, and a Gold Cup pushed $650. The worst news is: the prices are now even higher.

The good news may be Springfield's Model 1911-Al kit, a Series 70 with a suggested retail price of about $300 parkerized, $325 blued.

The kit comes in several sealed plastic bags.

After checking to see that no parts were missing, I scoured each with a vegetable brush and solvent, dried them, oiled everything generously, let them stand overnight, and wiped all down the next morning. It took me 30 minutes, working carefully, to complete assembly. (It might take you more or less time, but John Browning didn't design a nightmare. If you've never put one to-

This article first appeared in Rifle

gether before, the simplicity of the Model 1911 will amaze you, and any decent assembly manual makes the going easy.)

I had heard comments from critics regarding the Springfield's fit and finish. The kit I assembled went together precisely and was tight. There isn't a casting in the gun. The frame is forged and the slide is machined from bar stock that exceeds government specs for the "hard slide" military model. The barrel is hammer forged and broached, rather than button rifled. In other words, the Springfield is put together from good stuff. Except for factory markings and the absence of that famous rearing horse, the assembled Model 1911-A1 was "Old Slab Sides" in person. It looked pure GI and felt it with its 8½-pound trigger pull.

Initial test firing was conducted at 50 feet on

slow-fire targets. The ammo for the first few magazines was 230-grain military ball. Forty rounds were fired to break in the action. The combination of my hard-ball flinch, the heavy trigger, and sights these aging eyes could hardly see turned the best five-shot string into a group measuring 4 inches, low and left of the aiming point.

After switching to a lighter recoil spring, I fired several strings with a couple of target loads developed for my accurized .45s. I had to hand-feed the 215-grain rounds one at a time, but produced a group that included four in a surprising 1½ inches. The other light load tested, featuring a cast 200-grain round-nose, fed perfectly and resulted in a 2½-inch group. Point of impact, however, was still low and left. Some ball-and-dummy practice assured me I was not flinching. Low and left was where the gun wanted to print.

So far, it was apparent the Springfield was well-made in every respect, and, as expected, it wouldn't digest wadcutters.

Back on the workbench, I discovered something else. While sorting the fired brass, I found a nick on each case rim. That indicated the case was striking the upper right edge of the port as it was being ejected. The port would have to be opened up to correct the problem, a common one with stock. 45 autoloaders, and that was added to the list of things to do.

First, the recoil spring, guide rod, and recoil spring plug were replaced with a light weight spring, a full-length guide rod, and a modified spring plug, all of which are included in Wilson's Shok-Buff recoil system kit. Several poly fiber buffers, plus directions that make the installation kid stuff, are also included. The system really soaks up abuse otherwise intended for the shoot-

As delivered, Springfield Armory .45 (above) is precise image of John Browning's creation. Close inspection of kit's parts (right) reveals that Springfield Armory has not scrimped on quality.

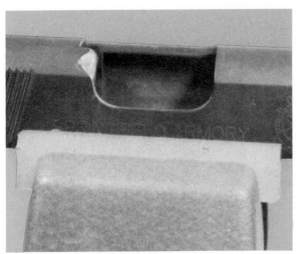

Ejection port is seen here after metal was removed from its upper rear. Area below port was hand-filed to ensure that its rim wouldn't interefere with ejected brass, and steel there was then polished and cold-blued.

ing arm, and is of great help in getting sights back on target after a round is on its way.

The next item was to get the 8-pound trigger-pull down to 3½ or 4 pounds and smooth out its rough spots. I poked around in the Brownell box of goodies and came up with a sear spring and a firing pin return spring. After installing the firing pin spring, tightly coiled end first, I polished the sear spring with crocus cloth where it contacts the sear and disconnector. The same treatment was given to all points of contact between sear and hammer, sear and disconnector, disconnector and trigger bow, trigger bow and frame.

A process of assemble, test, disassemble, polish, assemble, and test followed until the hammer fell at an even 4 pounds and that sensitive nerve about ½-inch in from the tip of everybody's trigger finger, including mine, announced that the rough spots had been polished away.

Next on the agenda was that displaced point of impact. Since I'd eliminated flinching as the cause, at least with target loads, two possibilities suggested themselves.

I checked out the first, the innards of the barrel itself, by forcing a soft lead .451 slug from breech to muzzle. As I drove it forward with mallet and dowel, resistance increased, which is exactly what should happen if the rifling is to have any influence on the bullet.

Next, my attention turned to the barrel—was it locking up out of line? With the slide closed, I eyeballed the relationship between the top edge of the barrel and the top edge of the slide back at the ejection port. The two edges were slightly

misaligned. The barrel was slanted muzzle down.

There are several ways to realign a barrel. A longer or shorter barrel link, depending on which way the barrel's tilting, is one. Welding and re-forming the barrel lugs is a second. Another is to modify the barrel link lug and barrel bushing. All are best left to an experienced gunsmith. My decision was to keep things simple and install Wichita's No. 203 cover-up rib.

Like all Wichitas, the 203 is machined from solid steel and has a V-groove milled into its fully adjustable rear sight. Somehow, that V emphasizes the front blade. The grooved rib itself is insulated, requires a minimum of machining to install, and weighs only 5 ounces. I've seen a bunch of Wichitas on silhouette and combat guns, but not many among bull's-eye pistols. Maybe some of us are a little slow latching on to a good thing.

Before the rib could be installed, I had to solve a problem discovered earlier: the matter of cases hitting the upper rear edge of the ejection port.

After wedging a fitted block of wood into the slide to prevent deforming its sides, I clamped the slide in a padded vise. The accompanying photographs show the area reworked—with carefully applied fine-tooth file and Arkansas stone—before and after the port was enlarged. The metal removed will vary from gun to gun, but the rear of the ejection port was widened by .005 inch and took away .015 rounding it off. After relieving the upper right edge as shown, I polished the result and touched it up with cold bluing. Firing a few rounds left no doubt that the

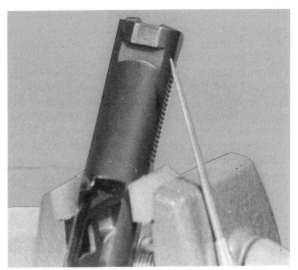

To prepare slide for installation of rib, front sight was removed and a flat was filed behind dovetail on rear of slide.

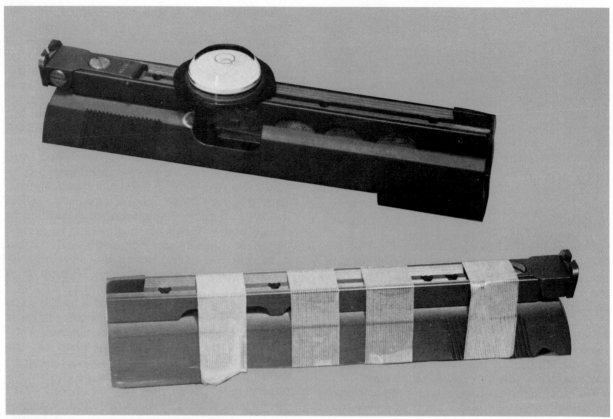

After coating underside of rib's rails with Acraglas and smearing release agent over top of slide, units were leveled (upper photo) and carefully mated. While Acraglas cured, slide was held in place with tape. Entire assembly was then placed on drill press and, using rib holes as guides, was drilled for 6-40 mounting screws.

case-nicking problem was a thing of the past.

To mount the rib, the slide was clamped between padded vice jaws and the rear sight was removed with the help of a brass drift (left to right). Two filing jobs followed. One on the front sight, which was lowered a little bit at a time with a ½-inch medium-cut mill file until there was a good fit between the rib and the slide's forward section. The other was between the dovetail and firing pin retainer, where the slide had to have a flat worked in to allow clearance for the rear sight area of the rib. The filed areas were then stoned smooth and cold-blued. After transferring the slide to the drill press table and truing it up in all directions, it was coated with Brownell's Acraglas release agent. The gel and hardener mix were applied to the rib's rails, then it was centered fore, aft, and in between, and finally taped in position to bed it to the slide.

Once the gel had cured, the taped and Acraglassed assembly was ready for drilling and tapping. Having determined the depth earlier, and using the rib holes as guides, four well-lubricated .095-inch-deep holes were drilled. I tapped them, flushed all bits of metal off the work with solvent,

secured the rib with four 6-40 mounting screws laced with Loctite, and removed the tape.

About now, eyebrows are being raised because nothing has been said or done about making the gun accept wadcutters. Why it hasn't is simple.

Springfield offers national match barrel and bushing sets to fit the Model 1911-A1 for an additional price of about $80. At the time I acquired my kit, none were available, so it arrived with a GI barrel. I subsequently decided to rework the Springfield only in ways almost anyone could duplicate. While mounting a rib might be among those ways, performing surgery on the throat of a GI barrel to allow it to digest wadcutters definitely is not.

Opening up a GI magazine, however, is. There's a little metal pimple on the surface of the magazine follower. I opened the lips of the magazine a little at a time from that point *back* until a case, when thumbed out of the loaded magazine, was released when it reached the pimple. This not only makes the magazine capable of feeding wadcutters, it makes it feed better generally.

Finally, I wrapped crocus cloth around a small

POST TUNE-UP ACCURACY

Springfield Armory Model 1911-A1

Bullet	Powder	Charge (grains)	Velocity (fps)	Largest 5-shot	Tightest 3-shot
				Group (inches)	
185 Winchester JWC	Bullseye	3.6	630	2.2	1.0
	231	5.0	665	3.4	.9
200 Hornady SWC	Bullseye	3.7	645	4.2	1.2
	231	4.5	640	3.3	1.6
	Herco	6.0	656	3.4	1.5
200 Sierra FMJ	231	5.0	690	3.9	1.8
	Unique	5.1	680	3.8	1.7
200 Cast RN	Bullseye	3.8	670	2.1	.5
	231	4.6	648	3.7	1.9
	Herco	5.5	600	2.9	1.4
230 Hornady FMJ	Unique	4.8	590	2.9	1.2
	231	4.5	598	1.7	.8
	Bullseye	3.8	610	2.1	.6

Tests conducted indoors. Ambient temperature, 34° F. Range, 25 yards; pistol supported by sandbag rest, with two-hand hold.

Bullets used in tests were: (1) 185-grain Winchester; (2) 200-grain Sierra; (3) 200-grain Hornady; (4) 200-grain cast Jones round-nose; (5) 230-grain Hornady.

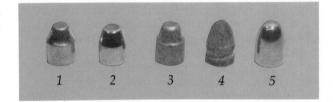

Final test targets reflect modified Springfield pistol's preference for Bullseye and round-nosed bullets. Jacketed 185-grain semi-wadcutters had to be loaded singly, but they punched out very creditable target groups.

triangular file and polished the rails of the slide and receiver before packing up my Skyscreen III and heading to the range for the final tests—the details of which are provided in the accompanying table.

The results of those tests were encouraging. Springfield's Model 1911-A1, as modified here, won't turn the average bull's-eye shooter into a champion or even assure a move up in class. (No gun and no amount of modification will.) What it does, and does very well, is make it possible for that shooter to own a solidly built, true-to-the-way-John-Browning-designed-it, .45 auto-loader that offers the potential of X-ring accuracy—and own it without going broke.

At today's prices, that sounds like a bargain.

Remedy for Ailing Accuracy

Bob Milek

We were squatted behind a colorless sandstone outcropping that capped a low ridge. Our varmint rifles were positioned in front of us, cradled on a variety of wheat-filled shot bags that we'd carried up from the truck. Before us the Wyoming prairie spread in an endless carpet of green, the expanse broken only by hundreds of small dirt mounds scattered randomly amid the low grass. The mounds were the telltale sign of a prairie dog town—burrows dug by the gregarious little rodents. From the fresh dirt evident on most of the mounds, this was a very active town.

"There's one," my partner grunted, instantly snugging the .22/250 varminter to his shoulder and closing the bolt on a live round.

I picked up the rodent in the field of view of my 10X40 Zeiss binocular. "He's 275, maybe 300 yards out," I counseled.

Seconds passed, then the .22/250 barked. A puff of dust erupted a couple of inches left of the dog, and I called the shot for my partner. He bore down and tried a second shot. This time it appeared that the dust kicked up just to the right of the prairie dog, which was standing erect, presenting a target some 10-12 inches high but no more than 3 inches wide. The third shot was

back on the left, but a bit too close for comfort. The varmint dived for the sanctuary of his burrow.

My partner groaned. "Guess I'm going to have to rebarrel this rifle. I've noticed that it's just not as accurate as it used to be. Seems like it gets worse all the time Probably shot out—I've fired maybe 3,000 rounds through it."

"When did you clean it last?" I asked. There was a long pause before the answer came. "Oh I guess it was last fall. But I've only fired 200 or 300 rounds through it since."

Only 200 or 300 rounds! No wonder his rifle's accuracy was steadily going to hell. I said as much and was quickly informed that in this day and age of noncorrosive primers and smokeless powder, regular bore cleaning was not only unnecessary, but for the most part was a waste of time.

True, we no longer have corrosive priming to deal with. However, that's only part of the picture. Powder fouling is a fact of life even today, but of even more concern is metal fouling. As a jacketed bullet passes through the bore, a fine film of powder residue and copper from the bul-

This article first appeared in Guns & Ammo

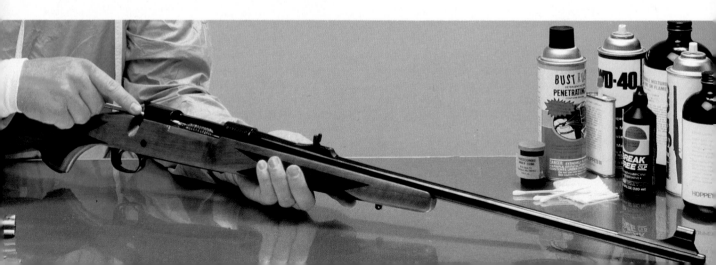

let jacket is deposited on the surface of the bore. Seldom are these deposits distributed evenly. Rather, the deposits are thicker in some places than in others. This, in turn, causes the next bullet passing through the bore to encounter varying degrees of drag-more where the fouling build-up is thickest, less where it's thinnest. The fouling continues to build as each successive bullet passes through the bore, becoming much heavier in places where the initial deposit was greatest. It follows, then, that as these deposits increase in thickness, the bullet encounters increased friction as well as increasingly worse "tight" spots in the bore. This affects velocity, barrel vibration, and bullet concentricity itself, resulting in continually decreasing accuracy as bore fouling steadily worsens.

Metal fouling builds less quickly and more evenly in a very smooth barrel than in a rough one, but it's a fact of life in all barrels—a fact that, if ignored, will destroy the accuracy of the finest barrel.

There's another kind of fouling encountered by shooters of metallic cartridges—lead fouling common to handguns in which lead-alloy bullets are fired. Lead fouling is deposited in much the same way as fouling from jacketed bullets, but becomes severe more rapidly and is much more difficult to remove. Why? Because the solvent compounds that attack gilding metal jacket fouling are almost totally ineffective on lead. As in rifle barrels, lead fouling is worse in a rough pistol barrel than in a smooth one, and the higher the velocity, the worse the fouling. A pistol bullet cast from a very hard alloy will lead the bore less quickly than one of a soft alloy, but if you think you have a lead bullet that doesn't lead the bore, you're kidding yourself. Some shooters labor under such a misconception because with most revolvers or semi-autos in which lead-alloy bullets are used, the accuracy is such that you won't notice the effects of leading until the condition is quite advanced. Leading occurs *anytime lead bullets are used*, thus steps should be taken to regularly remove lead deposits from the bore.

Bore cleaning can be overdone. If you've a .270 or a .30/06 through which you shoot 40-50 rounds a year, then I see little need to thoroughly clean the bore more often than after the hunting season. Likewise, if you shoot 300 or 400 rounds of jacketed-bullet ammo a season through your favorite .44 Magnum, don't panic if you clean it only at season's end. However, high-velocity varmint rifles; the new specialty pistols chambered for high-velocity rifle cartridges like the .223 Remington, 7mm Bench Rest Remington, etc.; and handguns in which either jacketed or lead-alloy bullets are used require regular, systematic cleaning in order to retain peak accuracy.

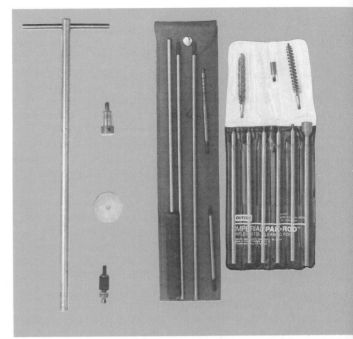

Lewis Lead Remover (left), is excellent for removing leading from bore. At right are two excellent sectioned cleaning rods for field use. Unit with longer sections is stainless rod from Belding and Mull; next to it is brass rod from Outers.

How you clean the bore is important. Done incorrectly, you may not remove the fouling, and if the wrong tools are used you can damage the bore. Let's talk first about cleaning rods. In any circle of shooters you can precipitate a hot argument simply over what material the rod should be made of—stainless steel, brass, or aluminum. As far as I'm concerned, stainless steel is best with good hard brass running a close second. I won't run an aluminum rod through the bore of any of my guns. Aluminum rods are soft, so they lack the strength required to force a tight-fitting patch through the bore without bending. And, because aluminum is soft, tiny particles of grit can become imbedded in the surface. When these rub the bore, they scratch it. No sir, I'll take stainless steel or good brass everytime.

Rod strength is an important consideration. Why? Because to thoroughly clean a bore, the patches must fit tight, and considerable pressure is required to force them through the barrel. If the rod bends, it will rub against the surface of the bore. If it breaks, you'll have a job getting the broken section on through without damaging the bore. Granted, quality stainless steel and brass cleaning rods are considerably more expensive than an aluminum rod but they're worth the money.

Should the cleaning rod be of one piece or in several short sections that screw together? One-piece rods are stronger, so I prefer them in my shop. They're also very difficult to pack into the field. Therefore, I think that you actually need two rods: a one-piece job for the shop and a sectional one to pack in your cleaning kit that goes to the field. Also, depending upon how many rifles of different caliber you must clean, several rods may be required. I normally use one rifle rod for .22 through .270 caliber and one of larger diameter for 7mm on up. Just one rod will suffice for large-bore handguns.

Of course, any rod is worthless unless the proper tip is used. Slotted tips are next to worthless because it's nearly impossible to achieve a tight patch fit in the bore. Far better are jag tips—and there's a variety available. Those offered by Outers for their brass and stainless steel rods are my favorites. They have a sharp point on which to spit a patch, and when the proper patch size is selected, these jags afford complete patch-to-bore contact. Another excellent jag is made by Parker Hale. This one is so constructed that paper toweling or cotton can be wrapped around the jag to achieve perfect fit. No patches are necessary. However, the Parker Hale jags I have fit only Parker Hale cleaning rods. They can be converted to fit American-made rods, but it's a job.

Bore brushes are also a necessity. I like good bronze brushes in preference to stainless steel, but this is just a personal thing. Likewise, I use a brush one caliber larger than the bore I'm clean-

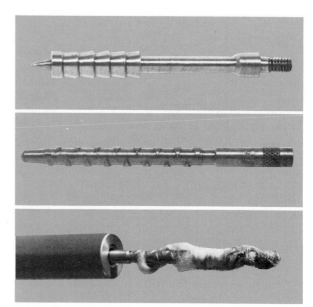

Jag tips are best choice for bore cleaning. When properly patched, they provide more patch/bore contact than slotted tips. Outers jag is at top; below it are Parker-Hale's tips; bottom one is wrapped with cotton.

ing—a 6mm brush for a .22, a .30-caliber brush for a 7mm, etc. In this way I achieve thorough brushing of the bore.

Now let's talk cleaning solvents. The variety out there is mind-boggling, and I wouldn't presume to tell you what to use. The important thing is that the solvent be one that attacks the fouling

Author made this cleaning kit especially for field use. Container is aluminum tool box that holds all necessary gear to keep Milek's guns in perfect condition in the field.

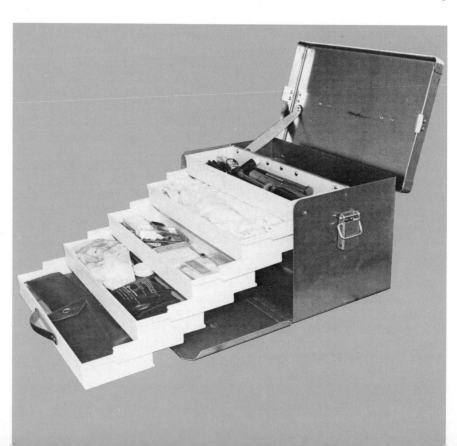

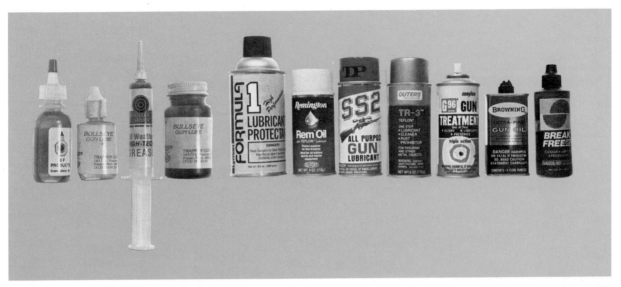

Shown in photo above left are four special lubricants. Lube from BF Products is formulated for stainless steel, while Bullseye lube works well on both stainless and blued steel. The other two, Shooter's Choice High-Tech grease and another Bullseye lube, are general-purpose products. Above right are just a few of popular lubricants.

you're trying to remove. For many years ammonia was the agent in bore cleaners that removed copper fouling. Today some compounds still contain ammonia and they work well. However, there are other chemicals that do the same thing and aren't as hard on wood finishes they may accidentally touch. The thing to keep in mind is that you need a solvent to remove both powder and copper fouling.

Which solvent do I use? Most of those available, I guess—which one at a specific time depending upon what's available and what I'm trying to do. My favorite for here at home where I have time to work is Shooter's Choice MC#7. This solvent is very effective at copper removal, yet poses a danger to the wood finish only if you let it stand on the wood for a period of time. MC#7 works a bit slower than some other popular solvents, so what I usually do is scrub the bore with a brush soaked in it, then run a couple of wet patches through the bore and allow the job to stand overnight. The next day a quick scrubbing with another wet brush and a few more patches, and all traces of jacket fouling are removed.

Sweet's bore cleaner is much harsher and faster-working, and you've got to be careful. Get it on the stock, and the finish is gone. Likewise, it'll take off bluing if allowed to stand on it. However, Sweet's works fast, particularly in a hot barrel, so it's my choice in the field. After I've fired for a while and the barrel is hot, I brush the bore with Sweet's, run a few wet patches through it, and presto—the copper fouling is gone.

Rig #44 is another bore cleaner that does the job.

Removing the copper fouling from badly neglected bores can be very time-consuming. It may take as many as 10 or 12 treatments with the best of solvents, and even then the job may not be complete. To speed things up, some shooters will opt to use a mild abrasive for tough jobs—J-B Compound by name. This is a paste laced with a very fine abrasive. There's always some controversy over whether or not to run an abrasive in a barrel, but I've heard no bad reports about J-B Compound. I'd certainly be judicious with its use, but I wouldn't hesitate in serious situations, particularly if it might help rejuvenate a barrel that would otherwise have to be replaced.

None of the solvents designed to remove copper fouling work on lead fouling. I know, many of the manufacturers claim their product does, but experience has shown me that this is just advertising. Solvents may remove lead in cases where very minute leading is present, but the conditions I encounter after a day of firing lead-alloy bullets usually call for more drastic action. The best system I know of is one where a special brass patch is stretched over an expandable rubber tip and pulled through the bore of a handgun. The tight-fitting patch scrubs all the lead out. The LEM Lewis Lead Remover was the pioneer in this area. This tool is still around in calibers .357, .41, .44, and .45, and Hoppe's offers an almost identical product.

After the bore is cleaned, the chamber and the bolt locking lug recesses in the receiver should

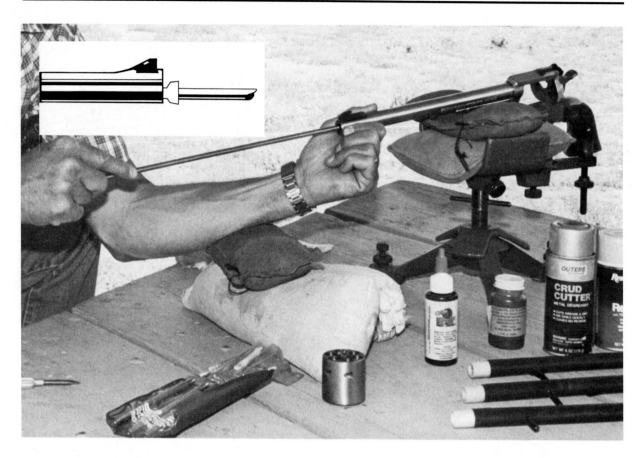

Revolvers require care in cleaning of bore and cylinder to prevent damage. Bore guide (above) protects muzzle from being damaged.

be swabbed clean. There are special tools for this, but a good job can be done with a large patch or piece of cloth on a cleaning rod.

Now consideration needs to be given to protecting the bore from moisture and rust. Only a good gun oil will do this. If you think there are a lot of bore solvents out there, just try cornering all the gun oils. It'll send you screaming down the street. Just about everything out there today is a combination lubricant/rust preventive, so use whichever ones you fancy. Do not over-lubricate the bore.

It's okay to oil the chamber to protect it when the gun isn't in use, but the chamber should be wiped dry with a degreasing solvent before shooting. When a cartridge is fired, the case expands to grab the chamber walls. If the chamber

is oily, the case fails to grab the walls, and all of the pressure is exerted rearward on the breech-face. The same thing is true in regard to the chambers in a revolver cylinder.

On bolt-action rifles or pistols, a dab of good gun grease on the rear surface of each locking lug' will greatly increase the ease with which the bolt closes and locks. This is the last step in the cleaning process, accomplished just as the bolt is being replaced in the rifle.

Finally, there's a right way and a wrong way to clean a bore. First of all, *never* clean from the muzzle end if you don't have to.

Second, always use a bore guide. This is simply a device that guides the cleaning rod, keeping it in alignment with the bore and preventing it from rubbing the chamber at the throat or rubbing

High-velocity pistols like this bolt-action should receive same care as rifle. Hornady's Ron Reiber "shuts down" frequently to clean bore of his pet Remington XP-100.

As part of cleaning operation, author degreases chamber and lubes rear surface of locking lugs for silky-smooth bolt operation.

MTM plastic bore guide is shown in use in pistol's receiver. Guide replaces bolt and not only guides cleaning rod but also prevents solvent from running back into chamber and magazine. Whenever possible, bore should be cleaned from breech to prevent damage to muzzle. Most autos, pumps, and lever guns, however, must be cleaned from muzzle.

against the bore itself. Cleaning rods, particularly stainless steel rods, are very hard and will actually cause wear if they rub the throat or the bore frequently. Bore guides come in many shapes and sizes, and even homemade ones work well. My favorites, though, are the plastic bore guides made by MTM Molded Products—those guys who make all the handy plastic shell boxes.

Finally, don't scrub a bore with a dirty patch.

Push the patch through, remove it at the muzzle, then withdraw the rod and use a clean patch for the next pass. In this way you never draw dirt and grit back into the bore.

As you can see, proper bore cleaning is essential to good barrel life and optimum accuracy. It's a tedious job, no doubt about that, and the proper equipment can get expensive. Nevertheless, if you neglect the bore of your rifle or handgun, you can't expect it to perform as it should.

ANNUAL UPDATE

Gun Developments

Jim Carmichel

Ordinarily, writing about new trends and discoveries in the field of sporting firearms is a fascinating chore, but in recent times I've felt somewhat like a sentimental jockey who is obliged to whip an aged racehorse around the course for yet another lap.

To be sure, there have been some encouraging new designs and even strokes of brilliance, but these have perhaps sparkled with more radiance than they warranted simply because of the dullness of the competition. What we saw taking place during the first half of the 1980s was the final act of a birth-and-death life cycle that characterizes the gun industry.

For the past half century and more I've kept a close eye on the gun industry, and have seen two such major cycles play themselves out. There are, of course, lesser cycles within major cycles, but as a rule we can pretty much count on the gun industry dying and being reborn every 12 to 15 years. Such cycles are common in nearly all types of business, but they are doubly devastating to the gun business because no one ever seems prepared for the lean years.

The low end of the most recent cycle occurred during a general economic recession, so it was easy, for a time, to blame sagging gun sales on hard times. But when times started getting better, gun sales didn't keep up with the general recovery. That's when a lot of gun sellers started wringing their hands and claiming the gun business was dead for all time. Some of the industry's trade shows I attended during this period were about as upbeat as a Greek tragedy. The sorry state of the gun business was blamed on everything except the guns themselves. "Nothing wrong with our guns," an executive was recorded as saying, "that a little dressing up won't fix." So for a couple of years we got old guns in new dress. But an outdated rifle with a racy new stock is like a faded actress with a new wig. She may turn your head for a moment, but soon you discover there's nothing new about her performance.

Though the catalogs of shooting history have been filled with truly wonderful firearms, the

This article first appeared in Outdoor Life

Top gun looks like a hump-back Browning, but it's really the new A-500 shotgun with a recoil-operated action. Second from top is the Browning Arms replica of the famed Winchester Model 71 rifle in .348 Winchester caliber, a great brush rifle. The third gun is Anschutz entrance-level Achiever target rifle in .22 Long Rifle. Spacers in stock permit changing length to fit the youngster. At bottom is the new Remington Model 11-87 auto shotgun, an improved Model 1100!

PHOTOGRAPHS BY TINA MUCCI

blunt truth is that exceedingly few guns have been blessed with enduring popularity. Being the fickle souls that we are, we simply get tired of the same old guns and yearn for something new and different. This general dissatisfaction with the same old guns is what triggers the bitter end of a sales cycle such as we've witnessed in recent years.

A case in point is Remington's Model 1100 autoloading shotgun. Introduced in 1963, it has been the most successful autoloader in history. A truly wonderful gun, it revolutionized not only the way we think about autoloading shotguns but also shotgun design in general. But there comes a time when the gun-buying public had to say "that's enough—the Model 1100 has been around so long we're tired of it. Show us something *different.*"

It's hard to put aside a proven winner, but there comes a time in every gunmaker's life when he has to realize that the only way to survive is with designs that are new and better and *different*

enough to capture the fancy of gun buyers. That's when the cycle begins anew. And that's where we are right now; the vintage of '87 is a milestone in sporting-firearms design.

This year guns are being introduced that will be the standards of comparison for the coming generation. Some, as you will discover, are dramatically new in concept and will be trend-setters. As a group, the vintage of '87 are the most interesting—and desirable—guns I've seen in a long while. They are mighty easy to write about.

BERETTA

When *Outdoor Life* Editor Clare Conley and myself, plus a few other American hotshots, traveled to England recently for the British Sporting Clays Championship, one of the things that most impressed me was the number of top British shooters who used Beretta shotguns. I didn't make a count, but it's fair to say that Berettas outnumbered all other brands combined. That says a lot about Beretta performance because those guys in England are mighty serious about their shooting.

Beretta very much intends to establish a similar leadership role in America, and to that end they are offering three new over/under shotguns to Sporting Clays. Keep in mind, by the way, that a good Sporting Clays gun is also just about the ultimate field gun because the game presents birds flying at all angles at varying distances.

The new Berettas are called the Models 686, 682, and 687L. All are 12-gauge, have 2¾-inch chambers, and 28-inch barrels with screw-in choke tubes. Each of the guns has a totally enclosed receiver that keeps the action free of powder residue and other trash. I've had only a quick look at these guns, and essentially the difference between the three seems to be a matter of wood, checkering, and engraving, with the 687L being the fanciest of the lot. All, of course, have such touches as single triggers, ventilated ribs, and deluxe finishing.

Beretta also has an upgraded version of its popular autoloader called the Model A-303. It comes in 12- and 20-gauge and features the famous Beretta gas system, which consists of only one moving part. The new A-303 handles all shell lengths and loads interchangeably without need for adjustment, and also features a magazine cutoff that allows hand feeding special loads into the magazine without emptying the magazine (such as dumping in a quick goose load when you are duck hunting).

The A-303 is a better-looking gun than the earlier autoloader, with nicer checkering. Naturally, the barrels are equipped with screw-in chokes.

Beretta U.S.A. Corp., 17601 Indian Head Hwy., Accokeek, MD 20607.

KIMBER

Remember the story about *The Little Train That Could*? Kimber is the little gun company that could—and does. Starting out with a stylish bolt-action .22 rimfire rifle, the good folks at Kimber (whose names are actually Warne) redesigned and improved their basic rifle until it has become the standard by which high-style rimfires are compared, replacing even the great Model 52 Winchester Sporter in the hearts of many aficionados. Then they turned their know-how to centerfire rifles and brought forth one of the sweetest miniature versions ever created, the Model 84. This year the Model 84 has been improved by the addition of a three-position safety that looks, feels, and works like a Model 70 Winchester safety. Also the Model 84 is now available with

a full-length Mannlicher-style stock called the Continental. And if you want to go the whole European route, the Model 84 is available in 5.6x50mm chambering.

Kimber has also entered the handgun field with a bolt-action varminter called the Predator. Available in two grades, the Kimber Predator is the best-looking of the space-age handguns and comes in .221 Fireball, .223 Remington, 6mm TCV, 6x47, and 7mm TCV calibers.

Kimber's really big news is the rifle thousands of hunters say they've been waiting for, a remake of the Model 70 Winchester as built before 1964. Called simply the Big Game Rifle, it is better looking than the old Model 70 ever was, by virtue of the classic-styled Kimber stock. The action, while not an exact duplicate of the original Model 70, retains all of its trim grace and style along with the Mauser-style extractor and three-position safety. Perhaps the best way to describe the new Kimber Big Game Rifle is as a Model 70 that has been customized by a topflight riflesmith. Available calibers are .270 Winchester, .280 Remington, 7mm Remington Magnum, .30/06, .300 Winchester Magnum, .338 Winchester Magnum, and .375 H&H Magnum.

This new rifle has too many features to describe here, but before long you'll be hearing all about what it has to offer. It is a truly great rifle that doesn't cost an arm and leg. I doubt if the company will be able to keep up with demand. I already have an order in.

Kimber of Oregon, Inc., 9039 S.E. Jannson Rd., Clackamas, OR 97015.

SKB

Remember SKB? This line of Japanese-made guns dropped out of sight a few years ago, presumably never to be seen again. But they are back, and given the promotional talents of their importer, Ernie Simmons III, we're going to be seeing a lot of SKB.

Until the decade of the 1960s, over/under shotguns were the ugly stepchild of the shotgunning family. Side-by-side traditionalists couldn't quite cotton to the odd stacked-barrel configuration, and the rest of us had trouble adjusting to the fancy prices charged for the few European-made over/unders available. These prejudices were overcome in a matter of months when the first Japanese-made over/unders reached these shores in the early '60s. These shotguns were surprisingly well made, ridiculously cheap, and shot as well as any gun on the market. Even smoothbore snobs couldn't resist their appeal, and before long the inexpensive imports were the mainstay of trap and skeet shooting. Many of the over/

unders were unabashed copies of the Browning Superposed shotgun, but one, the SKB, was distinctive in both design and feel. Its receiver, with a top bolt-locking system, was trimmer and more efficient. I can remember a time when fully 1/3 of the trap and skeet shooters at my club were equipped with SKB guns. And for the upland hunter, SKB offered some trim side-by-side doubles.

When SKB ceased production a few years ago, the guns were sincerely missed by a cadre of dedicated fans. Simmons tells me that SKB is coming back in a big way with a full line of sensational guns. I saw some samples a few months ago and couldn't help being impressed with them. The old SKB workmanship was there, but in fancier dress than ever. The thrust is toward high-performance equipment, and the new designs feel as good as they look.

Ernie Simmons Enterprises, 719 Highland Ave., Lancaster, PA 17603.

PARKER REPRODUCTIONS

If you've been crying the blues because you want to hunt waterfowl with a side-by-side double but are afraid of steel shot, you can dry your eyes. The people who bring us the beautiful Winchester Japan-built reproductions of the Parker double now offer a "Steel Shot Special." This special gun has thicker barrels than the standard model, plus the additional protection of chrome-lined bores. The Steel Shot Special comes in 12-gauge with 28-inch barrels choked Improved Cylinder and Modified. Chambering is for 3-inch Magnum, and the weight is a solid 7¼ pounds. The Steel Shot Special is being distributed by Jaeger's, Inc., a division of Dunn's, Inc., Hwy. 57E, P.O. Box 449, Grand Junction, TN 38039.

WINCHESTER–OLIN

Can you believe it? Winchester's Model 101 over/under shotgun was introduced 25 years ago. No wonder I'm beginning to feel old. I've shot a lot of game and a lot of targets with a lot of 101s over the past quarter century, and I never shot one I didn't like. The quality of workmanship in the 101 has, if anything, improved over the years, and it continues to be a lot of shotgun for the money.

The 101's 25th birthday is being celebrated with a fancy Silver Anniversary model of which only 101 will be built. With lots of engraving, fancy wood, and fancy checkering, the "one of 101," as it is called, will be a Winchester collector's delight.

For serious wingshooters, Winchester is introducing the Model 101 "American Flyer," a high-performance smoothbore designed to meet the tough demands of pigeon shooting. The American Flyer comes with two sets of barrels in 28- and 29½-inch lengths. The under barrel is fitted with a screw-in choke below a tightly choked top barrel. Pigeon shooters will appreciate this combination.

Now that the screw-in choke system is a way of life among shotgunners, many of us fail to give credit where it is due. In case you've forgotten, it was Winchester that pioneered this quick-change choke system back in about 1960. It was called the Versalite Choke at first and was fitted to Winchester's glass-barreled Model 59 autoloader. In 1969 the redesigned "Winchoke" became an optional accessory on some Winchester shotguns, and in time it brought about a revolution in shotgun chokes. Nowadays nearly every maker of shotguns offers some version of a screw-in choke, and all owe Winchester a vote of thanks for leading the way.

Beginning this year, the Winchoke has been redesigned so that the choke tubes fit flush with the muzzle. And to answer your first question, yes, the new flush-mount Winchoke tubes will fit your older Winchoke barrel. So see your dealer. Beginning this year, all Winchester "Classic Doubles" offered with the Winchoke feature will be equipped with the new Internal Winchoke System. A special fitting wrench comes with the new tube sets.

Olin Corp., P.O. Box 1355, 120 Long Ridge Rd., Stamford, CT 06904.

SIG-SAUER

The German-made Sig-Sauer Model 200 appears, at first glance, to be a more or less ordinary bolt-action rifle. When you take a closer look, however, you realize it is quite unlike any bolt rifle you've seen. First you'll notice that it has a two-piece stock. This feature is hardly unique; there have been two-piece stocks on bolt rifles before, but in this case both the butt section and forearm are easily and quickly removed with a single Allen-type wrench. With the butt section off, the rifle packs into a fitted carrying case that measures a scant 35½ x 12½ inches.

The next thing you'll discover about the Model 200 is that by loosening three screws on the under side of the receiver (with the same wrench you used on the stocks) the barrel can be removed. Another barrel in another caliber is fitted simply by tightening the three screws. It is one of the slickest barrel-swapping arrangements I've ever seen, but to be honest, I wasn't all that much impressed—yet. I've never been much of a fan of changeable-caliber rifles, because the ones I've tried have been too expensive, more trouble than they were worth, too complicated to be practical, woefully inaccurate, or, most often, combined all of these faults. In fact, the reason I agreed to test the Sig-Sauer Model 200 rifle was mainly to demonstrate to myself once again that quick-change, swap-barrel bolt rifles are, at best, only a gimmick.

Well, to make a long story short, I was pleasantly mistaken. The Sig-Sauer Model 200 proved to be one of the most accurate factory-made, hunting-weight rifles I've ever tested. Five-shot groups (100 yards) with a variety of .308 factory loads averaged close to an inch between the widest holes.

I changed to the .30/06 barrel, and the test results simply repeated themselves. The rifle seemed to have no bad habits; feeding from the clip magazine was smooth and flawless, and the bullets continued to punch into neat little clusters.

The Sig-Sauer Model 200 is a remarkable achievement in rifle design and engineering. There is a lot more to say about this rifle than my limited space allows, so the best thing for you to do is to take a first-hand look at your dealer's or to write for a brochure. This rifle is going to make a lot of experts rethink what they think they know about accuracy. I certainly changed

my mind about multiple-caliber bolt guns.

Sigarms, Inc., 8300 Old Courthouse Rd., Suite 885, Tyson's Corner, VA 22180.

ANSCHUTZ

The West German firm of Anschutz has become a legend because it makes the rifles that have won more Olympic and World Championship medals than all other makers combined. So when Anschutz introduces a new rifle, the shooting world takes notice. Usually it is a space-age marvel that pushes back the barriers of accuracy. This year, however, the company has looked into the future and seen that if target shooting—even

shooting itself—is to survive, we must begin with our youngsters. So the new Anschutz rifle is an entrance-level target model that "grows" with the child. It weighs only 5 pounds, so a small fry doesn't have to struggle with it, but at the same time it has a "big gun" feel. Spacers in the butt section permit the stock length to be adjusted to fit short-armed beginners.

Getting a kid in on an entrance-level target-shooting program can mean a lifetime of pleasure for him or her. All it takes to become an Olympic shooting champion is dedication and the right training. The new Anschutz rifle is a good piece of equipment for that all-important first step. Fittingly enough, this new training rifle is called "The Achiever."

Precision Sales International, P.O. Box 1776, 56 Southwick Rd., Westfield, MA 01086.

MARLIN

Marlin is like a stone bridge over troubled waters. Its products are tried and proven, it is proud of the guns it makes, and it introduces new models with cautious consideration. That's why Marlin guns don't tend to go out of style.

This year's new Marlin is called the "Midget Magnum." As the name suggests, it is a lightweight rifle chambered for the .22 Magnum Rimfire cartridge. More specifically, the "Midget Magnum" is a bolt-action, clip-fed (seven shots), take-down rifle that packs into a padded carrying case. The rifle is broken down for packing or compact storage by means of a single knurled screw in the forearm. The padded case not only protects the rifle for backpacking but even floats with the rifle inside. In addition to the adjustable open sights, the "Midget Magnum" also comes with a 4X scope, making it a "ready-to-go-shooting" rig.

Marlin Firearms, 100 Kenna Dr., North Haven, CT 06473.

The Smith & Wesson model 422 semi-auto (far left) is the handgun the company believes many women will want for self-defense. The rifles from left to right: High-grade version of the .30/30 Winchester is made by U.S.R.A.C. Note the checkering, luster finish and laminated stock. Shiloh Sharps replica is made in several big-bore chamberings. The new Baretta over/ under was designed for Sporting Clays.

WINCHESTER–USRAC

If you don't mind a minute sermon, let me tell you what's going on at U.S. Repeating Arms Company. During the past few years—in fact, ever since USRAC took over the historic Winchester factory in New Haven and continued production of Winchester Arms—it has been fashionable in some quarters to predict the firm's imminent demise. I've lost count of the times I've been tipped off by "industry insiders" that US-RAC couldn't make it another month. But the months have turned into years, and USRAC is still very much in the business of making good rifles and shotguns. Sure, it took its lumps with the rest of the industry during the recent recession, but the fact that it is still alive is proof that it's going to make it. At last count it had over 750 full-time employees busily making Winchester guns. I've decided that anyone who gets his kicks out of predicting the end of this historic old American firm would also enjoy seeing the Statue of Liberty topple into the muck.

In case you haven't been counting, this is the 50th anniversary of the Model 70 bolt-action rifle. That's right, 50 years ago Winchester brought forth what is widely considered the greatest sporting rifle ever conceived. It is the rifle that has it all—strength, accuracy, dependability, a stylish profile, and that undefinable something that makes a firearm great.

Naturally, USRAC is celebrating the occasion with a dolled-up limited-edition Model 70. Only 500 of this version will be built, and each will include fancy hand engraving, deluxe wood, and special numbering. Since the Model 70 is one of the most collected rifles of all time, I expect this special issue will sell out fast.

Of more interest to hunters is the addition of the .300 Weatherby Magnum to the list of calibers in which the Model 70 is offered. This bit of news is of more than passing interest because in addition to increasing the long-range reach and wallop of the Model 70, it is the first time a major U.S. armsmaker has offered a Weatherby Magnum caliber. Now we've come full circle because, as some old-timers will recall, some of Weatherby's first .300 Magnum rifles were built on Model 70 actions. One has to wonder what other Weatherby calibers will follow. I'll vote for the .257 Magnum; it's a wondrous thing. There are a few other modifications to the Model 70 line, mostly involving stock variations which include fiberglass and some striking stocks made of super-tough laminated wood.

A particularly nice item now offered by USRAC is a high-grade version of the old favorite Model 94 lever rifle in .30/30 caliber. The metal parts have a high-luster blue, and the stocking is in fancy wood with a deluxe fleur-de-lis checkering pattern. This item should appeal to lever-action fans who crave something nicer than the usual lever guns. USRAC is also offering a Model 94 in .30/30 with a 24-inch barrel.

For turkey hunters USRAC is offering a special version of the Model 1300 pump-action shotgun which sports a stubby 22-inch barrel and Win-Cam stocks. These stocks are of tough laminated wood in alternating shades of color for a pleasant camouflage effect. The ventilated-rib barrel is equipped with the Winchoke system, and a camouflage sling completes the package. For several years I've been touting short-barreled turkey guns. The idea makes a lot of sense.

U.S. Repeating Arms Co., 275 Winchester Ave., P.O. Box 30-300, New Haven, CT 06511.

STURM–RUGER

By the time you read this, Sturm-Ruger should be producing its new 9mm autoloading handgun, called the P-85. The model designation means it was to have been introduced in 1985 but that's not how things tend to work out at Ruger.

The introduction of this new 9mm autoloader is going to have considerable impact on the handgun market—for a number of reasons. First of all, it is a brilliant design with some engineering and design features that make it truly a gun of the 21st century. Second, it is remarkably easy to handle and shoot. I had a chance to test a prototype P-85 at Ruger's new Prescott, Arizona, plant, where it is being made, and found it hard to believe I was firing full-power 9mm loads. I did not have a chance to do any accuracy testing, but there are good reasons to expect exceptional accuracy from the P-85. My old pal Bill Atkinson, who is heading up P-85 production, is an accuracy buff himself and knows what it takes to make a gun shoot accurately.

The third reason the P-85 is going to make a big splash in handgun circles involves the U.S. Armed Forces' small-arms selection process. You'll recall the recent outrage when the U.S. government chose the Beretta 92-F as the standard sidearm. There has been a lot of debate over the fairness of the selection process, and as it turns out, new test trials will be held. Now that Ruger has a spanking-new high-tech 9mm to toss into test arena, the competitions should be mighty interesting indeed.

Sturm-Ruger, Inc., Southport, CT 06490.

REMINGTON

An awful fact is that there isn't enough room in these pages to describe all the new guns and model variations that Remington introduced in 1987. Remington unveiled its new line at a shooting-press bangfest recently, and I lost count when the tally of new items soared above the 100 mark.

The big news this year is an autoloading shotgun called the 11-87. I expect this new shotgun will be instantly regarded as a sort of "super Model 1100," which I suppose is natural enough considering the superficial similarities of the two guns. In fact, however, the new 11-87 is as far advanced beyond the M1100 as the M1100 was beyond its predecessor. The new 11-87, for example, which is initially available in 12-gauge only, handles any 12-gauge shell from the lightest

2¾-inch upland load to the heaviest 3-inch Magnum. No adjustments are necessary when changing from one load to another, thanks to a self-adjusting gas system.

The list of other new features built into the 11-87 reads like a list of solutions to common faults known to afflict the older Model 1100. For example, the extractor has been beefed up by nearly 1/3, the firing pin retractor spring has been redesigned, the magazine tube is now corrosion- and rust-resistant stainless steel, the feed latch attachment is stronger, the barrel support ring has been strengthened, the gas piston and piston seal are tougher, and the fore-end has been pinned to avoid splitting. And certainly not least, the gas mechanism is self-cleaning and *never* needs to be disassembled. Cosmetically, the new 11-87 is a considerably better looking and more stylish shotgun than the old M1100, but you need to see this for yourself and make your own judgments.

At present the 11-87 is offered in five basic forms: The "Premier" 11-87 is the standard field grade and comes with a ventilated-rib barrel in a choice of 26-, 28-, and 30-inch lengths. All barrels are equipped with the Rem Choke screw-in choke system.

The Model 11-87 Special Purpose version has dull-finished stock and metal surfaces for waterfowl and turkey hunting. Barrel lengths are 26 or 30 inches, all with Rem chokes.

The Model 11-87 Deer Gun also has a dull finish on metal and wood and, like the other SP model, comes with sling swivels and a camo sling. Barrel length is 21 inches and it is equipped with rifle sights and a special slug choke.

The 11-87 target grades for skeet and trap are chambered only for 2¾-inch shells. Both the trap and skeet versions are equipped with special Rem choke tubes.

According to the folks at Remington, the 11-87 is a wonderfully tough gun with enormous endurance. During one test, for example, 8,000 shells were fired without a malfunction. I've fired the 11-87 only a few hundred times, but already it is a favorite. It will surpass the Model 1100 as the greatest of the autoloaders.

Synthetic stocks for rifles are here to stay, and after a couple of years of pussy-footing around the issue, Remington has jumped in with both feet. Or should I say all three feet because the firm is offering a choice of three different synthetic stock materials. The two that you'll be seeing the most are the so-called "RS" and "FS" materials. The "RS" stands for "Rynite," a DuPont thermoplastic resin with 35 percent glass reinforcement. The coloring is either gray or camo, and the surface is slightly textured for a nonslip grip.

The "FS" stock is of the more traditional fiberglass and resin material reinforced with Kevlar. This is a stronger stock than the "RS" but is also more expensive. Of course, both the "RS" and "FS" stocks are totally waterproof and resistant to warpage due to moisture, dryness, and temperature changes.

Both the "RS" and "FS" stocks are available on Remington's Model 700 bolt-action rifle and are available as standard over-the-counter items.

An especially clever combination is the new Model Seven lightweight rifle with "FS" stock. The weight is only 5½ pounds, making this one of the fastest-handling rifles available. A second version of the Model Seven with synthetic stock is a special-order, custom-shop item that features a super-tough Kevlar stock and comes in .35 Remington and .350 Remington Magnum calibers. For close-cover bear hunting, I can't think of a rifle I'd rather use than this lightweight carbine-length bolt gun in the hard-hitting .350 Magnum chambering. It will also be great for black-timber elk hunting.

For accuracy buffs and long-range varmint hunters, Remington has introduced what is unquestionably the finest varmint rifle ever offered by a major arms maker. It is the 40-XB Varmint Special with a super-accurate Kevlar stock. And it is available—now, get ready—in .220 Swift chambering. That makes it the best of everything in a single package.

Another new Remington item that will appeal to rifle buffs is a Model 700 kit that comes with an unfinished stock. The metal parts are finished and blued and already inletted into the unfinished walnut stock. What you do is the final shaping and sanding, and apply the finish of your choice.

And at long last Remington offers a left-hand verson of its short-action Model 700 rifle. I know a lot of southpaws who will be happy. At present the choice in calibers is .243 and .308. We hope the short southpaw Model 700 will eventually include popular varmint rounds such as .22/250 and .223.

One more new item from Remington that will cause lots of grins is the Model 700 Classic in .338 Winchester Magnum chambering. The only sad thing about this combination is that it is a limited edition. A better combination for North American and African hunting is hard to imagine.

Remington Arms, DuPont, Inc., Wilmington, DE 19898.

BROWNING

The excitement at Browning is being caused mainly by its new A-500 autoloading shotgun.

Made in Belgium, this new self-loader will be especially fascinating to students of gun design because it is basically a recoil-operated mechanism at a time when gas-operated autoloaders are considered the design of the future.

Those who love the old square-backed Browning profile (and who doesn't) will feel right at home with the new A-500. The similarity, however, pretty much ends with the profile. Operating on a short-recoil system combined with a rotary bolt-locking mechanism, the A-500 is a unique blend of concepts. With the short-recoil system, the barrel, with the bolt locked in place, recoils into the receiver for a short distance (unlike the older long-recoil Auto-5, in which the barrel recoiled farther than the length of the shell). The rotary bolt then turns and unlocks and continues to the rear, extracting and ejecting the fired shell, while the barrel returns to its forward position. The claimed advantage to this system over a gas-operated mechanism is that there are no pistons or gas ports to foul and collect grime and residue.

The new A-500 fires any 12-gauge shell from the lightest to the heaviest loads without any adjustments and will feed and function 2¾- or 3-inch Magnum ammo interchangeably. The new A-500 comes in a choice of 26-, 28-, or 30-inch barrels, all with Invector Chokes. The fit and finish of this shotgun are typically Browning, which means that both wood and metal are beautifully done and show lots of hand finishing and fitting. The weight is about 7 pounds 5 ounces, which is a pound lighter than the old Browning auto in 12-gauge.

No one has been doing much to save the 16-gauge, but a couple of new shotguns from Browning might breathe some new life into the sentimental old favorite. One of these, the Sweet Sixteen autoloader, is hardly new, but it hasn't been available for a number of years. Now it's back, and as sweet as ever in the traditional square-backed Browning autoloader configuration. The new Sweet Sixteens even have the old round-knob grip like the early models. The only change is that the new Sweet Sixteens come with Invector Chokes, which is just the right amount of progress.

Browning's other new 16 is the classy Citori Over/Under. I've had several inquiries about 16-gauge over/unders of late, so I'm delighted that a really good one is available. The great thing about a 16-gauge double barrel is that it handles like a 20-gauge but performs almost like a 12. This new Citori weighs a scant 7 pounds with 28-inch barrels and is equipped with Invector Chokes.

Browning is offering big-game hunters some very interesting variations of the popular A-Bolt rifle. One of these is a left-hand version. It comes in the Medallion grade only and is available in .270, .30/06, and 7mm Remington Magnum calibers. Another A-Bolt variation, called the "Stainless Stalker," is made largely of rust-resistant stainless steel. The stock has a stippled, black paint finish which gives the rifle a rather racy silver-on-black look.

Another variation of the A-Bolt, called the "Camo Stalker," has a laminated wood stock with the laminations dyed various shades of green and tan. The effect is a natural camo pattern that is pleasing to look at. The metal parts of the Camo Stalker are dull-finished. Like the Stainless Stalker, the Camo Stalker is available in .270, .30/06, and 7mm Remington Magnum chambering.

If you have some gray in your hair, you probably remember Winchester's Model 71 lever-action rifle. It was a brute of a thing, but beautifully made and chambered for the frightful-looking .348 Winchester cartridge. I grew up in bear-hunting country, and a Model 71 was considered the boss of the woods. Production of the M71 was ended some 30 years ago, and since then it has become a much desired collector's item.

Now the Model 71 is risen from the grave in the form of a limited edition by Browning Arms—which is appropriate enough, seeing as how John Moses Browning himself was the gun's designer. (Actually, the Model 71 is a linear descendant of Browning's Model 1886 lever rifle).

Browning's new Model 71 can be had in both the "rifle" version with 24-inch barrel and the "carbine" style with 20-inch barrel. Also there are a couple of grades; one is the standard version with standard finish while the other has fancy wood and checkering plus a grayed-steel finish on the lever and receiver. This gray finish nicely sets off the engraving and gold inlay of the higher-grade version.

Browning Arms, Route One, Morgan, UT 84050.

MOSSBERG

Mossberg doesn't have anything radically new this year—which is understandable, seeing as how it introduced a whole new catalog of shotguns in '85 and '86. The folks there tell me that for the present they are making some minor changes with their existing line, such as developing better finishes, etc.

Anyway, they have a couple of items that are worth mentioning. One is the new Accu-Steel interchangeable choke tubes. These hardened stainless-steel tubes are especially designed for long-term use with steel shot. Rather than being of standard choke dimensions, they are tailored to deliver optimum patterns with specific sizes of steel shot. For instance, if you hunt geese with BBs or No. 1 or 2 shot, you can get a Full Choke Tube that delivers best patterns with the larger shot. But if you use 4s or 6s, there is another Accu-Steel tube that will deliver best Full Choke patterns. The choke and preferred shot size are marked on the tubes.

A gun that has been in Mossberg's catalog for a while but is often overlooked is the Model 1000 Junior autoloader. All too often, "youth"-sized shotguns are sawed-off economy models that don't have much to offer in the way of pride of ownership. The Mossberg Model 1000 Junior is a stylish autoloader that is very nicely fitted and finished. I've long considered the Model 1000 one of the smoothest-shooting and softest-kicking guns around. The Junior model comes in 20-gauge with a 22-inch Multichoke barrel. Actually, I'm not sure that "Junior" is the right thing to call this classy autoloader. It is just as good for ladies and compact-size gents.

O. F. Mossberg & Sons, Inc., 7 Grasso Ave., North Haven, CT 06473.

SHILOH SHARPS

Few names in firearms history conjure up such romantic images as does Sharps. These were the magnificent rifles that transported a westward-moving nation into the breechloading era. They were the king of the buffalo range and the queen of the target ranges. The huge Sharps cartridges were death and thunder wrapped in gleaming brass, and the man who owned a Sharps was someone to be reckoned with. The Sharps rifle company died with its namesake, and the rifles that survived that golden era are now treasured collector's items.

Over the past couple of decades there have been a series of efforts, here and in Europe, to build shootable replicas of the great Sharps rifles, but until now, none of the efforts gained any notable success—one reason being that the original Sharps were built almost totally by hand by some of the finest craftsmen of that time. Thus a truly good Sharps today would have to be built by hand with the same skill and loving care lavished on the originals.

The Sharps reproductions from the Shiloh Rifle Manufacturing Co. are built like the original—entirely by hand and virtually on a one-at-a-time basis. It's the kind of workmanship you have to see to appreciate.

Original Sharps rifles were built in a surprisingly wide variety of styles and calibers, and the good folks at Shiloh have set out to duplicate many of the originals. These range from the 1874 Military Musket to the Long Range Express sporting rifle. In between are the variations that made history. For traditionalists, Shiloh also offers the early 1863 models.

I've tested a couple of Shiloh Sharps, and they are wonderfully accurate. In fact, it would be a shame to own one just as a "looker," only because more than anything else they are built to shoot. The choice of calibers ranges from .32/40 up to the huge .50/100, with more modern calibers also available. I used a Shiloh Sharps in .40/70 with cast lead bullets to take a magnificent elk last season, and it was a truly memorable hunt. Write to Shiloh for its catalog and prices.

Shiloh Rifle Manufacturing Co., Inc., P.O. Box 279, Industrial Park, Big Timber, MT 59011.

SMITH & WESSON

Every year the firearms industry gets together to show its products to wholesalers and retail dealers. The huge affair is called the SHOT Show, and it serves as a coming-out debut for new guns and accessories. One of the hits of the 1987 SHOT Show was Smith & Wesson's new .22 rimfire autoloading pistol. It's called the Model 422, and its purpose is *fun*. That's right, just good old plinking and informal target potting for fun. I can't think of a better use for a .22 pistol. This one is engineered for safe, easy handling and accurate shooting.

Smith & Wesson makes no bones about the fact that the M422 was designed with women in mind. It sees women becoming increasingly interested in handguns for self-protection. "So, why not" the company reasoned, "offer a pistol that will also be a lot of fun to shoot?"

The M422 comes in two basic styles—one with checkered hardwood grips and adjustable sights, the other with fixed sights and grips of synthetic material. Both styles are available with either 4½- or 6-inch barrel lengths. The packaging includes a box of ammo and a bottle of oil to get you started off in a hurry. Pricewise the M422 is a pleasant surprise, as you'll discover.

From the design standpoint the M422 has some interesting departures, but most of the design features are centered around safety. S&W claims the M422 is one of the safest pistols on the market, and I see no reason to disagree. And there's no mistaking Smith & Wesson's look, feel, and workmanship. If any fault can be found, it is a stylistic one that can be best described by the Greek word *entasis*. Look it up and you'll see how it applies not only to Greek architecture but to the M422's grip shape and angle.

Smith & Wesson, 2100 Roosevelt, P.O. Box 2208, Springfield, MA 01102.

The Latest and Best Riflescopes

Jim Carmichel

The first fire of the winter warmed my favorite chair, and dancing light from the open gate shone through the decanter of unblended Scotch whiskey. I took a volume from a shelf of books too long unread. It had been written by a favorite gun writer. I had read it as a teen-ager, and as I turned the pages, certain passages were as well remembered as if I'd read them only the day before.

It was fascinating to note how little guns and ammunition had changed since the 1950s. The chapter on telescopic sights, though, was as outdated as a buyer's guide to horse-drawn vehicles. The improvements in scopes over the ensuing three decades can be fairly compared to the technological span from Lindbergh's solo flight across the Atlantic to manned space flight. If you go hunting next season with a new telescopic sight on your rifle or handgun, it will probably be more mechanically perfect and technologically advanced than any other piece of equipment you'll carry.

This is not to say that gunmakers cannot keep up with the pace. Developing a new firearm is a long and expensive process. Frequently, old models blend slowly into new ones. Scope makers, in contrast, are far more flexible, and can rapidly adjust to changing markets and new competition. A few years back, when the U.S. dollar was unusually strong against the world's other currencies, several European scope makers stepped up their sales efforts in this country. It seemed easy to take advantage of the favorable exchange rate. This exposed American shooters to an unprecedented array of high-quality European scopes at relatively affordable prices. In turn, this caused some confusion about quality among American shooters. Previously, the great house of Zeiss had represented the apex of German optical know-how. But then Americans were confronted by a line of prestigious scopes from

Swarovski, an Austrian firm, then another German line bearing the trademark Schmidt & Bender. In the meantime, the honored Austrian firm of Kahles was heard from again. These imported scopes carried regal price tags, and all four companies claimed to be without peers. What was a poor label-conscious scope buyer to do? For a while, American scope buyers tended to keep their money in their pockets, despite the exotic appeal of the European optics.

The problem was that European optics makers assumed that their scopes would also be well received in the U.S. After all, hunting is hunting, be it in the Bavarian forests or on the plains of Wyoming. Surely, American hunters would love their scopes, especially after becoming educated about the scopes' features. But to the immense surprise and bewilderment of the Europeans, it was they and not American buyers who had to be educated. The Europeans and their stateside advisers had failed to recognize basic differences between American and Continental hunting.

Big-game hunters in Central Europe typically shoot from a high seat or a blind, and most of the shooting is done at dawn or late dusk, when the game is not in deep cover. Therefore, European hunters have long favored scopes with enormous lenses that gather light. Such scopes are heavy and bulky, but this is of little consequence when shooting from a hoch sitz or a blind.

North American hunters tend to move a lot when hunting. We want our rifles and scopes to be portable and fast-handling. That's why the American scope has evolved into a comparatively compact unit, and it is also why American shooters and hunters didn't exactly beat down the door to buy the big European scopes with their pudgy tubes and bulging lenses. Clearly, if the

This article first appeared in Outdoor Life

Left to right: Schmidt & Bender 2½X-to-10X variable with big 56mm objective lens; Simmons 4X-to-12X variable-power Presidential-grade scope, a deluxe version; Bausch & Lomb Compact Model 2X-to-8X variable; Zeiss "Americanized" 4X C-series scope with standard one-inch tube; Burris 6X mini-scope with focusing objective for close-range shooting, especially with a .22 Rimfire; Redfield's 3X-to-9X compact; Weaver's new line includes this 3X-to-9X; Tasco 2X-to-7X variable with focusing objective lens; Weatherby's 3X-to-9X variable; 4X Nikon hunting scope; Leupold's fixed-power 8X with parallax set at 300 yards for long-range work; high-quality Austrian Swarovski 3X-to-9X is made in American style; 6X-to-24X Bausch & Lomb for varmints and serious target shooting. Wide variety of riflescopes makes it possible to choose one for a specialized form of shooting. In center is non-magnifying Aimpoint, which projects illuminated dot onto target or game.

Europeans were to compete successfully in North America, they would have to do so on American terms. Thus began the Americanization of some European scopes. Their makers had to shed glass like Gypsy Rose Lee shedding her feathers, and they had to abandon their beloved 26mm and 30mm scope tubes.

Zeiss led the way with its C-series scopes, which feature an Americanized profile and a 1-inch tube. This smaller tube diameter matched the preferred American mounting systems. Other firms have followed the Zeiss lead, and they are still introducing American-style scopes.

This is not to say that the big-name European scope makers took over the market. American firms came back tough and even gave the Europeans a taste of their own brew. For example, a couple of years ago, Leupold introduced a 6X scope that featured an oversize (by American standards) 42mm objective lens. This was Leupold's way of saying that if you like a big lens for shooting in dim light, you don't have to buy European to get it.

The European scope makers have a jealously guarded reputation for high quality, and this became a focal point of their advertising in American shooting magazines. There is little question that one of the primary goals of this quality campaign was to get a piece of the action being enjoyed by Japanese scope makers. During the 1970s, relatively inexpensive Japanese scopes become a major force in the U.S. market. Though most Japanese-made scopes are excellent, there were a few pretty bad ones, and these rotten apples cast a shadow on Japanese makes.

In true entrepreneurial spirit, however, some Japanese makers accepted the quality challenge and offered their own extra-high-quality scopes. An example of these efforts is Tasco's "World Class" line, which is sold with an unshakable warranty. Other Japanese-made scopes, such as Simmons' Presidential scopes, came forth on the quality front with such high-tech advances as one-piece tubes and superior lens coating.

Another interesting development during those revolutionary times was the introduction of big-name Japanese scopes. Nikon is synonymous with top-quality photographic equipment, so it was only natural for this highly respected optical company to enter the riflescope arena. Pentax did the same. The sales rationale of these heavy hitters in the optical world is that their world-wide recognition will virtually guarantee a rapid market response among hunters and shooters. I hear that these scopes are indeed selling well, and I suspect that the scopes themselves are excellent. I haven't tested the Pentax, but the two Nikons I've tried are superb.

Whatever a foreign scope maker's ambitions may be, however, he has to match the *usability* of scopes designed in the U.S. American scopes, some of which are made in Japan, represent the performance standard by which the world's scopes are judged, and American companies have also successfully maintained the technological cutting edge that others are forced to match. This is not just my personal judgment, but one that is candidly shared and often discussed by many executives of foreign optics companies. Yet, some foreign scope makers remain completely innocent of the facts as they exist in the United States.

During an average year, I receive three or four visits or calls from a foreign optics maker requesting that I try out a new riflescope and report on its potential. Almost invariably, there are follow-up calls that ask only about what I think of the optical qualities. When I report that the optics are excellent—which is usually the case—there are sounds of joy and happiness on the line.

"Oh, thank you very much, Mr. Carmichel. We appreciate your report."

"Hey, wait," I interrupt, "don't you want the rest of my report?"

Confused sounds on the line.

"What else, Mr. Carmichel?"

"The rest of the scope ain't worth a damn." Which is only true about half the time, but I like to get their attention.

More confused sounds on the line.

"But, you said the scope is excellent."

"No, I only said the optics are excellent."

From there on, the conversation often takes an ugly turn, but you get the idea. The dismal fact is that some foreign optics people don't know enough about shooting and hunting to be able to discuss the practical use of their own products. Some, especially the Asians, are so far removed from active shooting that potentially helpful feedback, if any, is ignored or short-circuited.

American designers, makers, and sellers of scopes have an enormous advantage because they are surrounded by shooting of all types. Even if they aren't active hunters and shooters themselves—most are—they often get more feedback than they can digest. When you consider this nonstop dialogue between experienced sportsmen and highly motivated scope makers, it's not difficult to discern why the American telescopic *sight* reigns supreme when it comes to getting a bullet on target. In other words, American scopes work because they work.

Foreign scope makers see themselves as optical firms that make telescopes; American scope makers think of themselves as makers of telescopic *rifle sights*. The gulf between these two viewpoints is enormous. American scope makers, including firms such as Bushnell that have U.S.-designed scopes built in Japan, view the product as a *mechanical* part of the rifle. It is not seen as an optical accessory used for looking at the target, but as a mechanical necessity for *hitting* the target. The scope must be a physical part of the rifle itself. That's why American scope-mounting systems are designed to rigidly lock scope and gun together. Many European mounts are of the quick-release type and do not lock up solidly every time.

During the past few years, the promotion and advertising of riflescopes, especially for foreign brands, has largely centered around qualities such as image brightness and others that are helpful or seem to be helpful when shooting at game during poor or almost nonexistent light. One could easily get the idea that a scope's brightness is its most important feature and that all game is bagged in near or almost total darkness.

Such promotions bother me somewhat. They tend to direct a potential customer's attention away from things that are of greater importance in scope performance, and, for that matter, away from what is important about hunting. The novice hunter could easily get the idea that all shooting is an early-dawn and late-dusk proposition, and not take into consideration the fact that many states specify legal shooting hours for big-game hunting. These laws, in effect, specifically prohibit shooting too early and too late, when the light is poor. Promotions that are built around low-light performance also bother me because they tend to ignore, and can cause customers to ignore, other, more important factors such as mechanical reliability and structural strength. In a list of such essential features as waterproofness, recoil resistance, adjustment stability, and adjustment repeatability, light-gathering capability ranks nowhere near the top when we are talking about American hunting. Besides, it is very difficult, if not impossible, for the average shooter to distinguish between the optical properties of most scopes. But the average shooter sure does know it if a scope fogs in wet weather, doesn't reliably maintain zero, causes the point of impact to shift when the power is changed (if it's a variable-power model), is too heavy, or has a finish that is not resistant to scratches and wear. These old-fashioned qualities largely determine if a scope has staying power in the market.

This doesn't mean that there have been no significant modern improvements. One of the most important has been the final coming of age of variable-power scopes. Jack O'Connor, who was shooting editor of OUTDOOR LIFE for many years, despised variable-power scopes, and rarely missed an opportunity to say so. He had good reason. One problem with early variable-power scopes was a tendency to shift zero when the magnification was changed. You could sight-in your rifle with the power setting, at, say, 9X, and then turn the power down to 3X and discover that the point of impact had shifted by as much as several inches at only 100 yards. Obviously, this was an intolerable situation. Old Jack loved to cuss about it, and the doctrine he spread cast a doubt on variable scopes that lingers to this day. Is that suspicion still justified? Leupold went a long way toward cancelling doubts about variable-power scopes a few years back with the introduction of its 6½X-to-20X Vari-X III model. With such a broad power range, significant zero shift seemed all but unavoidable. However, zero shift with this scope is virtually nonexistent. In fact, it was the first-ever variable-power scope to gain favor with the ultra-critical silhouette-shooting crowd. Other good variable-power scopes offer similar freedom from the zero-shift syndrome.

If you have doubts about any variable-power scope, you can test it without firing a shot by checking it with an optical collimator. These gadgets, which are normally used for sighting-in rifles, show you how much the crosshairs wander when the magnification is changed. Some gun shops will let you test scopes this way before buying, and allow you to use their collimators. I do not know of any shops, however, that will let you dunk a scope in the john for a test flush or two to check for leaks.

Here's a quick rundown on what's new or nearly new in the scope business. Some of the names may not be familiar, but they may loom very large on your hunting horizons.

AIMPOINT

This Swedish firm crept quietly onto the American shooting scene a few years ago with a non-magnifying optical sight that projects a battery-powered aiming dot onto the field of view. The scope's field of view is literally as big as all outdoors. It was, and is, an interesting and useful concept in aiming, especially at close ranges and when shooting at running game. The Aimpoint has really made its mark, however, in handgun competition. It has proven so successful in both combat and bull's-eye pistol shooting that four out of five winners use it. There are two new models this year, the 1000 and the 2000. The firm also offers mounting systems, and there are other accessories such as magnifying converters that add a bit of horsepower to a scope. I've shot 1-inch groups at 100 yards using Aimpoint-equipped rifles, so there's no doubt about accuracy.

Aimpoint U.S.A., 201 Elden St., Suite 302, Herndon, VA 22070.

BAUSCH & LOMB/BUSHNELL

Back in the 1950s, a hunter who used a Bausch & Lomb scope had the best sighting equipment money could buy. They were bright as diamonds, utterly waterproof, and tough enough to drive nails with, which is one way in which their dura-

bility was demonstrated. But riflescopes were only a small part of the company's huge optical line, and scopes were allowed to languish until the company bought Bushnell a few years ago. Now, the plan is to restore the B&L trademark to its former eminence among riflescope buyers. One way of proving what the company can do is to offer a super-sophisticated scope to the highly critical target-shooting trade. This is an area where very few scope makers dare to tread, but B&L is offering high-powered glass in 24X, 36X, and a 6X-24X variable. I've given these scopes a tough workout, and performance is dazzling.. For hunters, B&L offers a full line of riflescopes and handgun scopes. They are fairly pricey, but reek of quality. I don't mind paying for excellent performance.

The Bushnell line is more complete than ever, and reflects the Bausch & Lomb determination to be the major force in riflescopes and handgun scopes. To get an idea of how much quality and *usability* you can get for your money, take a look at Bushnell's Sportview 4X or 10X scopes for rimfire rifles and air rifles.

Bushnell Division of Bausch & Lomb, 300 N. Lone Hill Ave., San Dimas, CA 91773.

BURRIS

Burris, like its founder, Don Burris, is largely content to let its products speak for themselves. And speak very well they do. Don Burris is an unsung hero of scope design with a number of revolutionary patents to his credit. Before founding his own company, he was the guiding genius at Redfield. The Burris scopes reflect Don's know-how, and have earned a solid reputation for innovation, dependability, and performance—the real stuff of American scope making. New from Burris this year is a line of Silhouette scopes in 10x, 12X, and 6X-18X variable that feature target-style knobs and focusing objective lenses. Burris handgun scopes are much praised by competitive shooters, and I expect their Silhouette scopes will win similar approval. Also new are shotgun scopes in 1½X and 2½X. These are tough scopes, built to take the battering of shotgun recoil, and they have a huge field of view. The 1½X model, for example, has a 62-foot field at 100 yards.

Also new is a very, very sweet 6X mini-size scope that features a focusing objective lens. This means it can be focused for close ranges for best performance with rimfire rifles or air guns. Despite its compact size, it is a grown-up scope in every way. This scope gets my vote as one of the most notable shooting items of the year.

Burris Co., 331 E. 8th St., Greeley, CO 80632.

LEUPOLD

This is the 40th year for Leupold & Stevens, and the company has lots to celebrate about. Mostly, it can celebrate the fact that it set the standard by which other scopes are judged. It achieved this recognition by offering no-nonsense riflescopes and handgun scopes that get the job done under the toughest conditions. Any scope maker who claims that his product is as good as a Leupold is saying a mouthful and will be hard-pressed to back it up. In competition-shooting circles, Leupold scopes are simply a fact of life.

Leupold's new scope for this year seems modest, but it may be quietly revolutionary. It is a trim 8X riflescope that is almost identical to Leupold's popular 6X hunting scope. This makes the 8X great for long-range hunting for such game as pronghorns, mule deer, and bean-field whitetails. The focusing is preset at 300 yards, so you won't have to worry about parallax on long shots.

Leupold & Stevens, Inc., Box 688, Beaverton, OR 97005.

NIKON

A name with the magic of Nikon cannot be ignored, but the company isn't resting on its laurels. Every lens in a Nikon scope is coated, and the image quality and brightness are dazzling. If you consider the price of a Nikon scope in the light of its quality, you'll discover that the value is unbeatable. Nikon has an excellent service backup and a growing list of dealers. I expect that the company is here to stay.

Nikon currently offers four scopes that effectively cover the hunting end of the shooting market. These include a fixed-power 4X model and variables in 1½X-4½X, 2X-7X, and 3X-9X magnifications.

Given Nikon's optical know-how and their considerable facilities for testing and development, they could very well redefine what we think we know about the precise aiming of rifles. If they offer some scopes in the upper magnification ranges, we'll see how good they really are and how serious they are about the shooting market.

Nikon, Inc., Sport Optics Dept., 623 Stewart Ave., Garden City, NY 11530.

PENTAX

Like Nikon, Pentax is a riflescope that enters the market preceded by a great reputation for camera optics. I have not tested any Pentax scope, so I cannot say anything about performance, except to remark that Pentax is a great name to uphold. At present, the company offers 4X and 6X fixed-power models plus 2X-7X and 3X-9X variables. There is also a 3X-9X compact model.

Pentax Corp., 35 Inverness Dr., East Englewood, CO 80112.

REDFIELD

The name Redfield is synonymous with fine performance. Over the years, the company introduced revolutionary concepts in scope design. For example, it can be credited for the design and introduction of the first truly satisfactory variable-power scope. Everyone who makes a variable scope today owes Redfield a vote of thanks. For years, the company virtually owned the target-shooting and varmint-hunting market by virtue of its unbeatable 3200 and 6400 scopes. In recent years, though, Redfield quietly exited from this part of the market to concentrate on making solidly dependable hunting scopes. Rumor has it that the company may be once again aiming at the competitive shooting market. If so, thousands of precision shooters who grew up with Redfield will rejoice at the news.

New from Redfield this year is a 1X-4X variable in the Golden Five Star line. Earlier in 1987, I tested a Five Star 3X-9X scope, and it is a tough act to follow. The Five Star scopes have one-piece tubes plus a lifetime warranty.

While we're on the subject of Redfield, have you ever looked through one of the Illuminator scopes? It's quite an experience. The folks at Redfield are good people who know what hunting is all about.

Redfield Gun Sight Co., 5800 E. Jewell Ave., Denver, CO 80224.

SCHMIDT & BENDER

If you are hooked on German scopes and German quality, you'll find the Schmidt & Bender line irresistible. The company represents old-fashioned Teutonic quality nicely married to modern technology. The prices are rather spectacular, but the guys I know who use Schmidt & Bender scopes swear they won't use anything else. A pal of mine who dotes on hard-kicking African rifles tells me they are the only scopes that seem utterly resistant to very heavy recoil. The company's most popular scope, I'm told, is the huge 2½X-10X variable with a 56mm objective lens. Farm boys in the Deep South love this scope for late-evening shots at whitetails.

Schmidt & Bender scopes are available from Paul Jaeger, Inc., a division of Dunn's, Inc., Box 449, 1 Madison Ave., Grand Junction, TN 38039.

SIMMONS

This sparkling new optical firm is the brainchild of Ernie Simmons III, third generation of a famous name in American shooting products such as the well-known Simmons shotgun rib. Ernie knows hunting, and he knows telescopic sights. His goal is to provide the latest technical advances in reasonably priced high-performance scopes. For example, Simmons scopes combine one-piece-tube construction with the latest in lens coatings. The results are tough, hard-working scopes at workingmen's prices. The looks, finish, and packaging are top quality. I recently tested a Simmons Presidential Model in 4X-12X configuration, and my conclusion was: "Who could ask for anything more in that kind of scope?" There are dozens of scopes in the Simmons line, far too many to describe here, so see your dealer.

Simmons Outdoor Corp., 14205 S.W. 119th Ave., Miami, FL 33186.

SWAROVSKI

Though relatively new on the American scene, Swarovski is one of the grand names in Austrian optics and glassmaking. Those magnificent chandaliers you see in many great European houses and hotels, for example, are often Swarovski products. The first Swarovski scope I tried was something of a hybrid between a battleship and an astronomical telescope. It was a wonderful sight, but it was too heavy to carry very far. Now, Swarovski offers an Americanized line of scopes that I rank among the very best and most usable. I hunted with the 4X model a couple of years ago, and found its performance utterly faultless.

Swarovski Optik, 1 Kenney Dr., Cranston, RI 02920.

TASCO

The Tasco name has become one of the best known in the scope field. The various models in the Tasco catalog are far too many to list here, so see your dealer. If you can't find a Tasco scope for your kind of hunting at a price you want to pay, you just aren't trying. As with some other made-in-Japan scopes, the company is trying hard to beat the rap by offering high-quality, high-performance scopes backed by an unbeatable warranty.

Tasco Sales, Inc., 7600 N.W. 26th St., Miami, FL 33122.

WEATHERBY

It's not exactly a secret that Weatherby has been offering scopes nearly as long as the company has been making high-performance rifles. The quality is good, and the dollar value is excellent. Next time you look at a Weatherby rifle, take a look through a Weatherby scope as well. You'll be pleasantly surprised.

Weatherby, Inc., 2781 Firestone Blvd., Southgate, CA 90280.

WEAVER

Weaver is the grand old name in riflescopes. Weaver first brought telescopic sighting to the average American hunter and revolutionized the way we aim our rifles at game. When the firm failed a few years back, it was a bitter loss to the industry, but now Weaver's back. The outfit is now a division of Omark, the huge company that also makes Speer bullets, CCI ammo, RCBS reloading tools, and Outers gun-cleaning and refinishing supplies. Omark is run by shooters and hunters, so you can count on the dependability. There are nine models in the Weaver line at present.

Weaver Division of Omark Industries, Inc., Box 856, Lewiston, ID 83501.

ZEISS

The optical world snaps to attention when the name Zeiss is mentioned. Commenting on the quality of this honored optical firm would be redundant. If Zeiss isn't convinced what it's making is the best, the company won't make it.

Zeiss was perhaps the first firm to offer German-built American-style scopes—its well-received C-series. This year, the C-style 3X-9X scope has been redesigned to eliminate some of the mounting problems inherent in the earlier model. Also, the company is offering European-style Z-series scopes. Unlike the C-series scopes, which have standard 1-inch tubes, the Z-series scopes have the traditional 26mm tubes in the fixed-power models (26mm is a bit more than 1 inch) and 30mm tubes in the variable models. The 1½X-6X variable has some particularly interesting possibilities. Other models include a 2½X-10X variable and 4X, 6X, and 8X fixed-power scopes. The 8X model has a great big 56mm objective lens.

Zeiss Optical, Inc., 1015 Commerce St., Petersburg, VA 23803.

Index